Quantum mechanics sets fundamental limits on the amount of information one can extract from a system with a single set of measurements. Recent results of new theoretical analyses and sophisticated optical experiments have approached these limits as never before, giving rise to a more complete knowledge of the quantum properties of light. This book is the first to give a detailed description of this fascinating branch of quantum optics.

The book is self-contained and begins with a description of some key results and tools from quantum optics, dealing particularly with the quantum theory of light and the concept of quasiprobability distributions. The author discusses the quantum mechanical description of simple optical instruments before giving a detailed treatment of quantum tomography (optical homodyne tomography). The book concludes with a chapter devoted to the problem of the simultaneous measurement of position and momentum.

Graduate students and researchers in quantum optics will find this a fascinating book. It will also appeal to anyone interested in the foundations of quantum mechanics or more general problems of quantum measurement.

T0268993

CAMBRIDGE STUDIES IN MODERN OPTICS

Series Editors

P.L. KNIGHT

Department of Physics, Imperial College of Science, Technology and Medicine

A. MILLER

Department of Physics and Astronomy, University of St Andrews

Measuring the Quantum State of Light

TITLES IN PRINT IN THIS SERIES

Measuring the Quantum State of Light

ULF LEONHARDT

University of Ulm

CAMBRIDGE UNIVERSITY PRESS
Cambridge, New York, Melbourne, Madrid, Cape Town, Singapore, São Paulo

Cambridge University Press
The Edinburgh Building, Cambridge CB2 2RU, UK

Published in the United States of America by Cambridge University Press, New York

www.cambridge.org
Information on this title: www.cambridge.org/9780521497305

First published 1997
This digitally printed first paperback version 2005

A catalogue record for this publication is available from the British Library

Library of Congress Cataloguing in Publication data

Leonhardt, Ulf, 1965-
Measuring the quantum state of light / Ulf Leonhardt.
p.   cm. – (Cambridge studies in modern optics)
Includes bibliographical references and index.
ISBN 0-521-49730-2
1. Optical tomography.   2. Quantum optics – Technique.
I. Title.   II. Series.
QC449.5.L46   1997   96-44409
535 – dc21   CIP

ISBN-13 978-0-521-49730-5 hardback
ISBN-10 0-521-49730-2 hardback

ISBN-13 978-0-521-02352-8 paperback
ISBN-10 0-521-02352-1 paperback

Dedicated to Harry Paul

Contents

1

Introduction

1.1 A note to the reader

The writing of *Measuring the Quantum State of Light* was quite a challenge and an exciting adventure. This field is rapidly developing, new ideas are appearing and revising previous work, and yet a book is supposed to be ageless and continuously timely. One challenge was to finish the calculations in time for finding the "classic" solutions of a number of problems in this area. This book contains these new and, hopefully, long-lasting results. Because the field is growing and diversifying, I was also forced to select the most developed material and to focus on just two classic paradigms of state measurement – *quantum tomography* and *simultaneous measurement of position and momentum*. Quantum tomography (optical homodyne tomography), especially, is a relatively simple and highly efficient experimental technique for investigating the quantum properties of light. This book may help to propagate this remarkable scheme so that it can become an experimental standard in quantum optics. The second paradigm, the simultaneous measurement of position and momentum (via eight-port homodyne detection) demonstrates many theoretically intriguing features of the quantum nature of light. Apart from being practical experimental techniques, both schemes may stimulate our conceptual understanding of quantum states. These experiments show how quantum phenomena are occuring in the real world (and not in the Hilbert space only).

Another challenge was that I had three types of readers in mind while writing this book. One is the expert who is actively doing research in quantum optics. This reader probably will be mostly interested in the technical details examined in the core of this book, in the chapter "Quantum Tomography." I hope that he or she will find there the necessary know-how for doing or understanding quantum-state reconstructions. Another potential reader is the graduate student who is going to do research in this area. The student should actively read this book with paper and pencil from the beginning to the end. He or she will find a brief

1

introduction to the quantum theory of light, a detailed survey on quasiprobability distributions, and a chapter on various aspects of simple optical instruments. Wherever possible, I tried to focus on the discussion of the physics and not on the formal mathematical problems. Many short derivations are explained only in words. Students are welcome to reproduce these calculations (because this is an excellent way of learning the subject). A solid basic knowledge of textbook quantum mechanics and some experience with the underlying mathematics are the only requirements. A third type of reader is the generally interested physicist who "wants to know everything about quantum mechanics (but never dared to ask)." I hope this reader will learn at least something about alternative formulations of quantum theory (via quasiprobability distributions and in terms of observable quantities), the wave-particle dualism, and the puzzling nature of vacuum noise, to name a few examples. What I enjoyed most while writing the book are the various connections of the single theme "measuring the quantum state of light" to quite a number of exciting quantum effects.* The challenge was to inform the expert, to teach the student, and to entertain the generally interested physicist. At least I tried to combine these "classically contradicting" complementary features. Let us begin with some questions.

1.2 Questions

Since the golden age of quantum mechanics, most physicists have more or less accepted that quantum objects are rather abstract. Quantum states are Hilbert-space vectors (or statistical ensembles of vectors), and Hermitian operators are recruited to describe physical quantities. Why is that necessary? Why are "quantum things" not just "real things"; why are they abstract and unfamiliar? Maybe because the "real" things we see do not happen to be quantum objects. Are we simply much too macroscopic to be comfortable with atoms or elementary particles? Suppose we could magnify the quantum world by a subtle apparatus. What would we see? Could we then see the "quantum things"? Not quite! Seeing quantum objects means disturbing them, in general. Seeing quantum objects from all points of view so that we can learn what they really are means disturbing them with certainty. The overall back-action of observations cannot be reduced much below Planck's constant. We cannot see the things *as they are* because as soon as we watch them they behave differently. Instead, we see only the various aspects of the physical objects, such as the wave or

---

*Although this book does not really go "Vom Himmel durch die Welt zur Hölle" [107] it touches some speculations, goes through solid quantum optics, and, finally, comes down to measurement technology.

particle aspect, that depend on the particular kind of observation. Moreover, these features are complementary; they exclude each other, and yet they are only different sides of the same coin. So, probably, we must accept that quantum objects are fundamentally different from familiar "real things" and that their true nature cannot be seen in a single experimental setting.

Suppose, however, we prepare quantum objects repeatedly in identical states and observe their complementary features in a series of distinct experiments. Each experimental setting probes one particular aspect. Could we then put all the pieces of the puzzle together to infer what the state of these objects is? Can we reconstruct pictures of "quantum things" from a complete set of observations? How would these pictures look? And what happens if we attempt to measure the complementary aspects in a single experiment? In this book we essay an answer by studying two paradigms of state determination.

SOBALD DIE ATOMIS BEOBACHTET WERDEN,
BENEHMEN SIE SICH AUF EINMAL GANZ ANDERS

Fig. 1.1. "As soon as we watch the atoms they behave differently." Quantum objects are significantly changed by measurements, in general. They are showing us only their particular aspects, the wave or the particle aspect, for instance, but not what they really are. [Reproduced with the friendly permission of P. Evers.]

The first example is quantum tomography. Classical tomography is a method for building up a picture of a hidden object using various observations from different angles. Computer-assisted tomography, for instance, gives insight into a living body by evaluating recorded transmission profiles of radiation that has penetrated the body from various directions. In quantum optics, tomography has been applied experimentally to reconstruct the quantum state of light from a complete set of measured quantities. These observables comprise all complementary aspects a light beam may have. We will study the detection scheme and the mathematical and physical background to understand how this remarkable experiment works. We will use the so-called quasiprobability distributions to picture quantum states in a classical fashion. We will see how these distributions manage to combine the complementary features of quantum systems, and we will examine their properties. Finally, we will use quasiprobability distributions to show how quantum states can be tomographically reconstructed from experimental data.

What happens if we dare to measure position *and* momentum simultaneously? We will try to answer this question by analyzing a second paradigm of state measurement. We will study an intriguing device for a joint yet imprecise measurement of canonically conjugate quantities. In this experiment we do see an overall picture of the quantum object "light," but the picture is fuzzy. The two paradigms are methods to measure the quantum state of light, that is, to gain as much information as possible about light.

Why light and not other quantum objects? Light is a wonderful object to perform experiments with. Lasers can generate light of superb quality, optical devices can process light with great precision, and highly efficient detectors are available to measure the quantum properties of light. Classical optics is a well-established century-old theory, and so we understand very well what the classical features of light are and can focus on the nonclassical quantum effects. This is the reason that many fundamental tests of quantum mechanics have been performed in quantum optics. (Remember that the very history of quantum theory began with Planck's radiation law.) So quantum optics has much to offer to those who are interested in practical demonstrations of fundamental quantum principles. Moreover, light is the most likely candidate for practical applications of state measurements. Light is a typical high-technology tool to investigate or to change various properties of matter. By gaining as much information as possible about light, we can better explain the behavior of material probes. Additionally, because light is used for communication, certainly worth studying is how to extract the maximal information allowed by the very principles of quantum mechanics. Measuring the quantum state of light could be an important issue for fundamental questions and practical applications as well.

1.3 Quantum states

Before we begin, let us take a step back and remember what we are usually doing in physics. Imagine a typical quantum-optical experiment (similar to the schemes described later in this book). A master laser generates a train of light pulses. They are processed on an optical table and guided to a crystal (with nonlinear optical properties), where they generate light pulses of a different optical frequency. These pulses are the objects to investigate. They are carefully protected from any disturbances or losses and directed to a detection device for measuring their physical properties. As is the case in most experiments, three steps are involved in the procedure. The first process is the preparation of a physical object. In our example the object is the light pulse generated in the crystal. All produced pulses should be identical to guarantee reproducible results. After the preparation, the object is protected from the environment and evolves in a controlled way. Finally, some physical properties of the pulse are measured. The experiment is repeated on each pulse of the train to eliminate statistical errors. Of course, this procedure presupposes that the prepared physical objects are indeed identical. They have lost their individuality and are regarded simply as samples in a series of experiments. Note that there might be some uncontrollable fluctuations involved in the preparation process. The phases of the master pulses may be random, and their intensities may vary. We assume, however, that the averages of the measured quantities tend to certain values when the experiment has been repeated sufficiently often. The observed facts should be at least statistically reproducible. This assumption must be carefully checked. Quite typically, significant effort is required to obtain reproducible results that can be considered true features of the prepared objects. If the physical objects are statistically reproducible, then they are regarded as members of a *statistical ensemble*.

The separation of identical objects from the rest of the world is the key assumption of physics. The objects should differ only in their *states*. We may vary, for instance, the amplitudes or the phases of the prepared pulses, but we still regard them as light pulses. Knowing the state means knowing the maximally available statistical information about all physical quantities of a physical object. Physical theory describes mathematically how the observable quantities are related to one another and to the state. For this description the object is mirrored in a mathematical model that should be as simple as possible yet still in accordance with the observed facts. The theory itself repeats mentally the steps of a physical experiment. Creating a model means simplifying and separating ideas from each other and assigning them to physical quantities. Then mathematics is employed to process the abstract ideas, and finally they are retranslated into physical terms to predict the measured quantities.

1.3.1 Classical physics

Classical physics assumes that in principle we could perfectly separate physical objects from the rest of the world. These objects should behave completely predictably when they are tested in physical experiments. According to classical physics we could see the things without disturbing them. To quote Poincaré [220] –

"We have become absolute determinists, and even those who want to reserve the rights of human free will let determinism reign undividedly in the inorganic world at least. Every phenomenon, however minute, has a cause; and a mind infinitely powerful, infinitely well-informed about the laws of nature, could have foreseen it from the beginning of the centuries. If such a mind existed, we could not play with it at any game of chance; we should lose.

In fact for it the word chance would not have any meaning, or rather there would be no chance. It is because of our weakness and our ignorance that the word has a meaning for us. And, even without going beyond our feeble humanity, what is chance for the ignorant is not chance for the scientist. Chance is only the measure of our ignorance. Fortuitous phenomena are, by definition, those whose laws we do not know.

But is this definition altogether satisfactory? When the first Chaldean shepherds followed with their eyes the movements of the stars, they knew not as yet the laws of astronomy; would they have dreamed of saying that the stars move at random? If a modern physicist studies a new phenomenon, and if he discovers its law Tuesday, would he have said Monday that this phenomenon was fortuitous? Moreover, do we not often invoke what Bertrand calls the laws of chance, to predict a phenomenon? For example, in the kinetic theory of gases we obtain the known laws of Mariotte and of Gay-Lussac by means of the hypothesis that the velocities of the molecules of gas vary irregularly, that is to say at random. All physicists will agree that the observable laws would be much less simple if the velocities were ruled by any simple elementary law whatsoever, if the molecules were, as we say, *organized*, if they were subject to some discipline. It is due to chance, that is to say, to our ignorance, that we can draw our conclusions; and then if the word chance is simply synonymous with ignorance what does it mean? Must we therefore translate as follows?

'You ask me to predict for you the phenomena about to happen. If, unluckily, I knew the laws of these phenomena I could make the prediction only by inextricable calculations and would have to renounce attempting to answer you; but as I have the good fortune not to know them, I will answer you at once. And what is most surprising, my answer will be right.' "

Chance is ignorance. However, as Poincaré also pointed out, even a minute lack of knowledge about the initial conditions of a nonlinear system may lead to completely unpredictable phenomena or, in modern terminology, to chaos. This behavior explains the success of statistical methods in the foundations of thermodynamics and the seemingly final complete victory of classical physics immediately before the dawn of the quantum era.

The physical objects of classical point mechanics are mass points moving in empty space and being subject to forces. The state of an individual object is characterized by the position q and the momentum p. But even if we do not know q and p precisely for the individuals of an ensemble, we can still characterize the total ensemble by a state, as long as we observe at least statistically reproducible facts. In this case q and p fluctuate statistically according to a certain probability distribution, $W(q, p)$. This distribution $W(q, p)$, called a *phase-space density*, represents the state of the ensemble of mass points. It describes the maximal statistical information we have. In a classical field theory, such as electrodynamics or general relativity, the space itself is assumed to be a physical object, a field, and the field state at every space-time point is characterized by a field strength or, more generally, by a statistical distribution of possible field strengths (which are compatible with the field equations).

In any case, the state of an object in classical physics is a physical property itself. In principle the state could be observed without inducing a disturbance. No fundamental obstacle exists to eliminating all statistical fluctuations from an ensemble of physical objects because such fluctuations are assumed to be caused entirely by our lack of precise knowledge.

1.3.2 Quantum mechanics

However, this presumption of classical physics was proven wrong in the era of quantum mechanics. Even when the preparation of a single physical object is optimally under control, no guarantee exists that all physical properties of this object are predictable. However, repeated measurements on identical objects still show that the statistical frequencies of physical quantities converge to fixed values. In this statistical sense, the physical properties of identically prepared objects are reproducible and we can still describe an ensemble of physical objects by a state. Theoretical prediction is possible yet restricted to the calculation of probabilities for events to occur.

What is the reason for this intrinsic statistical uncertainty? As already mentioned, observations disturb quantum objects in general. Observations cause an uncontrollable back-action of the object onto the rest of the world, and vice versa. Consequently, the object behaves unpredictably. Suppose, however, that we repeat one measurement immediately after it has been performed and that the physical object has not been destroyed in the measurement process. Under ideal circumstances we would read the same measurement value as before. So the first experiment itself prepares the physical object in a state that is perfectly matched by the particular experimental setting. In this state the object causes no back-action and behaves predictably, indicating that cases exist where the physical object, being in a given state, is not influenced by measurement. On the other hand, the observation of one particular feature disturbs other potential aspects the object might have, aspects that are complementary to the observed quantity. According to Heisenberg's uncertainty principle, for instance, we cannot measure position and momentum simultaneously *and* precisely. While observing the position of a mechanical system we are losing the momentum information. We cannot see the things as they are. They might rather resemble abstract ideas than things we call visible and real. What we do see are only the different aspects of a quantum object, the "quantum shadows" in the sense of Plato's famous parable [219]; Plato compared people to prisoners who were chained in a cave and forced to see only the shadows of the things outside and not the things as they are.

Finally we remark that quantum mechanics is a rather universal theory that describes our approach to diverse physical objects such as elementary particles, nuclei, atoms, light, or semiconductor excitations, to name just a few examples. Paradoxically, this universality may also mean that quantum mechanics describes our universal way of doing physics rather than the universe itself. However, the basic assumptions of quantum mechanics are not likely to be reduced to pure logic; they are, indeed, assumptions, and consequently they contain nontrivial information about us and the physical world.

1.3.3 Axioms

Let us recall the basic axioms of quantum theory, and let us try to motivate them. This book is of course not the place for a comprehensive development of the theory. We assume that the reader is already familiar with the basic formalism of quantum mechanics. However, because some of the ideas touched in this book illustrate fundamental issues of quantum physics, we would find it appropriate to turn "back to the roots of quantum mechanics" in a brief and certainly incomplete survey. Moreover, some of the presented arguments will be explicitly used later in this book. Let us sketch, in a couple of lines, *one* possible way of motivating the principal ideas of quantum theory.

"At the heart of quantum mechanics lies the *superposition principle* – to quote from the first chapter of Dirac's classic treatise [78] '. . . any two or more states may be superposed to give a new state' " [243]. We denote the state of a perfectly prepared quantum object by $|\psi\rangle$. Then, according to this principle, the complex superposition $c_1|\psi_1\rangle + c_2|\psi_2\rangle$ of two states $|\psi_1\rangle$ and $|\psi_2\rangle$ is a possible state as well. In other words, perfectly prepared states, called *pure states*, are vectors in a complex space. The superposition principle alone does not make physical predictions, it only prepares the ground for quantum mechanics.

Nevertheless, the principle is highly nontrivial and can hardly be derived or taken for granted. In the history of quantum mechanics the superposition principle was motivated by the wavelike interference of material particles. Note, however, that this simple principle experienced a dramatic generalization such that we cannot consider its historical origin as a physical motivation anymore.

Let us now turn to more physical assumptions. When we observe a physical quantity of an ensemble of equally prepared states, we obtain certain measurement values a (real numbers) with probabilities p_a. Given a result a, we assume that we would obtain the same result if we repeated the experiment immediately after the first measurement (provided, of course, that the physical object has not been destroyed). This assumption is certainly plausible. As a consequence, the object must have jumped into a state $|a\rangle$, called an *eigenstate*, which gives the measurement result with certainty, an event called the *collapse of the state vector*. Or, if we prefer to assign states only to ensembles of objects, a measurement produces a statistical ensemble of states $|a\rangle$ with probabilities p_a. According to the superposition principle we can expand the state vector $|\psi\rangle$ before the measurement in terms of the eigenstates $|a\rangle$, written as $|\psi\rangle = \sum_a \langle a \mid \psi\rangle |a\rangle$, with some complex numbers denoted by the symbol $\langle a \mid \psi\rangle$. What is the probability for the transition from $|\psi\rangle$ to a particular $|a\rangle$? Clearly, the larger the $\langle a \mid \psi\rangle$ component is (compared to all other components) the larger should be p_a. However, this component is a complex number in general. So the simplest possible expression for the transition probability is the ratio

$$p_a = \frac{|\langle a \mid \psi\rangle|^2}{\langle \psi \mid \psi\rangle}. \tag{1.1}$$

Here $\langle \psi \mid \psi\rangle$ abbreviates simply the sum of all $|\langle a \mid \psi\rangle|^2$ values. It is a special case of the more general symbol

$$\langle \psi' \mid \psi\rangle = \sum_a \langle \psi' \mid a\rangle\langle a \mid \psi\rangle \tag{1.2}$$

with the convention

$$\langle \psi \mid a\rangle = \langle a \mid \psi\rangle^*. \tag{1.3}$$

The mathematical construction (1.2) of the symbol $\langle \psi' \mid \psi\rangle$ fulfills all requirements of a scalar product in a vector space. However, at this stage the scalar product depends critically on a particular set of eigenstates $|a\rangle$ or, in other words, on a particular experiment. Let us assume that all possible sets of physical eigenstates form the same scalar product so that no experimental setting is favored or discriminated against in principle. This assumption seems to be natural yet is highly nontrivial. If we accept this, then the symbol $\langle \psi' \mid \psi\rangle$ describes *the* scalar product in the linear state space. We can employ Dirac's convenient bra-ket formalism, and in particular we can understand the $\langle a \mid \psi\rangle$ components as orthogonal projections of the $|\psi\rangle$ vector onto the eigenstates $|a\rangle$.

Formula (1.1) is the key axiom of quantum mechanics. It makes a quantitative prediction about an event in physical reality (the occurrence of the measurement result a), and it contains implicitly the superposition principle for describing quantum states. The historical origin of this fundamental principle is Born's probability interpretation of the modulus square of the Schrödinger wave function.

Now we are in the position to reproduce the basic formalism of quantum mechanics. Because the probability p_a does not depend on the normalization of the state vector $|\psi\rangle$, we may simplify formula (1.1) by considering only normalized states, that is, we set

$$\langle \psi \mid \psi \rangle = 1. \tag{1.4}$$

Because the eigenstates yield the measurement result a with certainty, they must be *orthonormal*,

$$\langle a \mid a' \rangle = \delta_{a'a}. \tag{1.5}$$

Furthermore, the system of eigenvectors must be complete,

$$\sum_a |a\rangle\langle a| = 1 \tag{1.6}$$

if we assume that any observation yields at least one of the values a so that $\sum_a p_a = \langle\psi| \sum_a |a\rangle\langle a||\psi\rangle$ equals unity for all states $|\psi\rangle$. The average $\langle A \rangle$ of the measurement values a is given by

$$\langle A \rangle = \sum_a a p_a = \langle\psi|\hat{A}|\psi\rangle \tag{1.7}$$

where we have introduced the Hermitian *operator*

$$\hat{A} = \sum_a a|a\rangle\langle a| \tag{1.8}$$

with *eigenvalues* a and *eigenvectors* $|a\rangle$. [The structure (1.8) explains the term *eigenvectors* for the measurement-produced states $|a\rangle$.]

We must mention another fundamental axiom of quantum mechanics concerning the composition of physical objects. If one system consists of, say, two subsystems, then the theory should allow us to experiment on each of the subsystems independently. We would obtain two real measurement values (a_1, a_2) and if we had repeated the same experiment immediately after the first measurement we would read the same values (a_1, a_2). Furthermore, we would also obtain a_1 if we had performed the repeated measurement only on the first subsystem, irrespective of what happens on the other (irrespective of which measurement is performed there) and, of course, vice versa. So it is natural to assume that independent measurements correspond to factorized eigenstates

$$|a_1, a_2\rangle = |a_1\rangle \otimes |a_2\rangle. \tag{1.9}$$

As usual, the symbol $\otimes$ denotes the tensor product. Note, however, that this innocent-looking axiom is capable of peculiar physical effects when it is combined with the superposition principle. The state space of the total system is the tensor product of the subspaces. However, the superposition of two different states $|a_1\rangle \otimes |a_2\rangle$ and $|a_1'\rangle \otimes |a_2'\rangle$ will not factorize anymore in general, producing an *entangled state*.

The total system is not a mere composition of its parts, because the subsystems are correlated. This correlation may bridge space and time, showing the potential nonlocality of quantum mechanics, as expressed for instance in the Einstein–Podolsky–Rosen paradox [84] and in Bell's inequalities [26], [27].

As we have seen, we can reproduce the basic mathematical machinery of quantum mechanics starting from some ideas about states and measurements. These ideas have been distilled and formulated quantitatively in axiom (1.1). However, the basic formalism is far too general to be sufficient for solving specific physical problems. Here we rely on physically motivated guessing to find the significant physical quantities and their relations within the general framework of quantum mechanics. In particular, we need this physical information to understand the classic quantum effects such as the quantization of the energy (or of other observables).

1.3.4 General quantum states

So far we have considered only pure states, presupposing a perfectly controlled preparation of physical objects. It is, however, not difficult to relax this assumption and to extend the concept of quantum states to ensembles of physical states as long as the very idea of reproducible physical objects does make sense. For this extension we assume that we have at least statistical information about the prepared states, that is, we have an ensemble of pure states $|\psi_n\rangle$ with probabilities ρ_n. The prediction $\langle A \rangle$ of any physical quantity must be the average of the expectation values $\langle \psi_n | \hat{A} | \psi_n \rangle$ for the individual states $|\psi_n\rangle$ with respect to the preparation probabilities ρ_n, or

$$\langle A \rangle = \sum_n \rho_n \langle \psi_n | \hat{A} | \psi_n \rangle$$

$$= \sum_a \sum_n \rho_n \langle \psi_n | \hat{A} | a \rangle \langle a | \psi_n \rangle$$

$$= \sum_a \langle a | \sum_n \rho_n | \psi_n \rangle \langle \psi_n | \hat{A} | a \rangle. \tag{1.10}$$

We write the last line in terms of the trace

$$\langle A \rangle = \text{tr}\{\hat{\rho} \hat{A}\} \tag{1.11}$$

introducing the *density operator* (sometimes also called the *state operator* [11])

$$\hat{\rho} = \sum_n \rho_n |\psi_n\rangle\langle\psi_n|. \tag{1.12}$$

The representation of $\hat{\rho}$ in a given basis is called the *density matrix*. We interpret the density operator (1.12) as the most general description of a quantum state and the formula (1.11) as the general rule of predicting observable quantities. Pure states are of course included in this general concept because their density operators are projectors $|\psi\rangle\langle\psi|$. States that are not pure are called *mixed states*.

We may also use the concept of density operators to generalize our fundamental axiom (1.1) about quantum measurements to broader circumstances. According to formula (1.1) the occurrence of a measurement result a is associated with a "jump" of the pure state $|\psi\rangle$ to the pure eigenstate $|a\rangle$. The probability for this process is given by the scalar product $|\langle a \mid \psi\rangle|^2$ (assuming the normalization of $|\psi\rangle$). Suppose that we are not completely certain about the state $|\psi_n\rangle$ and about the particular measurement result a but that we can still describe statistically the state as well as the observation. In this case the probability $p(A)$ of the measurement A is given by

$$p(A) = \sum_a \rho_a \sum_n \rho_n |\langle a \mid \psi_n\rangle|^2 = \text{tr}\{\hat{\rho}_a\hat{\rho}\} \tag{1.13}$$

where we have introduced the density operator

$$\hat{\rho}_a = \sum_a \rho_a |a\rangle\langle a| \tag{1.14}$$

for the eigenstates $|a\rangle$ occurring with probabilities ρ_a. This theory of states and measurements is assumed to be valid as long as the very notion of statistically reproducible facts is appropriate.

What can we say about density operators in general? First of all, $\hat{\rho}$ is Hermitian and normalized,

$$\text{tr}\{\hat{\rho}\} = \sum_n \rho_n \text{tr}\{|\psi_n\rangle\langle\psi_n|\} = \sum_n \rho_n \langle\psi_n \mid \psi_n\rangle = 1, \tag{1.15}$$

because the individual states and the probability distribution ρ_n are normalized. The density operator is strictly *nonnegative*, that is, it has only nonnegative eigenvalues, because for all $|\psi\rangle$

$$\langle\psi|\hat{\rho}|\psi\rangle = \sum_n \rho_n |\langle\psi \mid \psi_n\rangle|^2 \geq 0. \tag{1.16}$$

Note that this obvious constraint may be very difficult to handle. Given a mathematically constructed operator $\hat{\rho}$, we cannot decide easily in general whether $\hat{\rho}$ meets the physical criterion (1.16). Representing $\hat{\rho}$ in the eigenbasis, the eigenvalues of $\hat{\rho}$ can be interpreted as probabilities (because they must be normalized and nonnegative) for the eigenstates. Consequently, any normalized

Hermitian operator can be accepted as describing a quantum state as long as the operator is nonnegative. Note that the unraveling of a mixed density operator in terms (1.12) of an ensemble of individual pure states is not unique if these states are not orthogonal to each other. For mixed states there is no unique way of telling whether statistical fluctuations of observed quantities are caused by fluctuations in the state preparation (by our subjective lack of knowledge) or by fluctuations caused by the measurement process (by our fundamental lack of complete control).

How can we discriminate pure from mixed states or, more generally, characterize the purity of a state? One option is the von-Neumann entropy

$$S \equiv -\text{tr}\{\hat{\rho} \ln \hat{\rho}\}. \tag{1.17}$$

The entropy vanishes for pure states only, exceeds zero for mixed states, and, most important, is an extensive quantity for nonentangled subsystems $\hat{\rho}_1 \otimes \hat{\rho}_2$ because in this case

$$S = S_1 + S_2. \tag{1.18}$$

The von-Neumann entropy is regarded as the fundamental measure of preparation impurity for quantum states. However, the entropy might be difficult to calculate. Another computationally more convenient option is the *purity* $\text{tr}\{\hat{\rho}^2\}$ or the *purity parameter*

$$S^{pur} = 1 - \text{tr}\{\hat{\rho}^2\}. \tag{1.19}$$

Using the eigenbasis of the density operator, we see that

$$\text{tr}\{\hat{\rho}^2\} = \sum_n \rho_n^2 \leq \sum_n \rho_n = 1. \tag{1.20}$$

The equality sign holds only for pure states, and the purity $\text{tr}\{\hat{\rho}^2\}$ thus discriminates uniquely between mixed and pure states. Because $1 - \rho_n$ is less than or equal to $-\ln \rho_n$ for $0 < \rho_n \leq 1$, the purity parameter gives a lower bound

$$S^{pur} \leq S \tag{1.21}$$

for the von-Neumann entropy S.

What happens if we have a composite system and observe only quantities A_1 referring to one subsystem? In this case we can simplify the general rule (1.11) for predicting A_1, introducing the *reduced density operator*

$$\hat{\rho}_1 = \text{tr}_2\{\hat{\rho}\}, \tag{1.22}$$

where the trace tr_2 should be calculated with respect to the unobserved degrees of freedom. The expectation value $\langle A_1 \rangle$ is given by

$$\langle A_1 \rangle = \text{tr}_1\{\hat{\rho}_1 \hat{A}_1\}, \tag{1.23}$$

"tracing" here only in the observed subsystem. The so-constructed reduced operator $\hat{\rho}_1$ obeys all requirements for a physically meaningful density operator – it is normalized, Hermitian, and nonnegative because the total density operator meets these criteria. Consequently, we can regard $\hat{\rho}_1$ as describing the quantum state of the reduced system. The parts of a composite system are genuine quantum objects, being in mixed states in general. In this way the theory itself shows that we can separate a single object from a larger system and describe it by a density operator. Quantum mechanics is consistent with the a priori assumption of separable physical objects. (In fact, the explanation why matter occurs in stable and identical atomic units has been one of the most significant achievements of quantum theory.) Note that even if the total system is in a pure state, the reduced system might be statistically mixed. This intriguing feature relies on the entanglement of the subsystems (and hence it can be used as a measure for entanglement [20], [21]). We cannot observe all aspects of an entangled system by considering the subsystems only. Our lack of knowledge about the partner object causes statistical uncertainty in the state of the subsystem, explaining why the reduced system may be in a mixed state.

1.3.5 Remarks

We have associated the key elements of quantum mechanics with certain structures in Hilbert space. The *state* of a quantum object is described by a Hermitian, normalized, and nonnegative density operator $\hat{\rho}$. The observable features of the object are associated with Hermitian operators $\hat{A}$ in the sense that the expectation value of a physical quantity is given by the trace formula (1.11). Finally, subsystems are composed by forming the tensor product of the substate spaces. (Later, we will sketch alternative formulations of quantum mechanics based on quasiprobability distributions or on directly observable quantities. See Chapter 3 and Section 5.1.4. Note, however, that these forms have subtle intrinsic problems. See the discussion in Sections 3.1.2 and 5.1.4.)

The mathematical structure of quantum mechanics is clear and simple, yet the physical interpretation is still a subject of considerable debate, and so is the interpretation of quantum states. To quote Ballentine [11] –

The concept of *state* is one of the most subtle and controversial concepts in quantum mechanics. In classical mechanics the word state is used to refer to the coordinates and momenta of an individual system. Since it has always been the goal of physics to give an objective realistic description of the world, it might seem that this goal is most easily achieved by interpreting the quantum state function (state operator, state vector, or wave function) as an element of reality in the same sense as the electromagnetic field is an element of reality. Such ideas are very common in the literature, more often appearing as implicit unanalyzed assumptions than as explicitly formulated arguments.

According to Ballentine, "the assumption that a quantum state is a property of an individual physical system leads to contradictions." This book is not the place to settle the debate about the physical interpretation of states in quantum mechanics. The book would already fulfill one of its objectives very well if it could stimulate the discussion about the nature of quantum states by showing practical examples of experimental quantum-state reconstruction. We will not dive deeply into philosophical debates but let, if possible, the physics speak for itself.

1.4 Further reading

A look at the history of the quantum-state reconstruction may be interesting. W. Pauli [212] raised the question of whether the Schrödinger wave function $\psi(q)$ can be inferred from position and momentum distributions, that is, from $|\psi(q)|^2$ and $|\tilde{\psi}(p)|^2$. The general problem of state reconstruction was stated by U. Fano in his classic article [90] on density matrices.

This problem and the Pauli problem were treated by W. Gale, E. Guth, and G.T. Trammell [99]. Ambiguities in the original Pauli problem were shown by J.V. Corbett and C.A. Hurst [63]. They pointed out that if $\psi(q)$ has a definite parity, then both $\psi(q)$ and the complex conjugate $\psi(q)^*$ lead to identical momentum distributions $|\tilde{\psi}(p)|^2$, as can be easily verified. Consequently, the Pauli problem has not a unique solution. Nevertheless, R.W. Gerchberg and W.O. Saxton [102], [103] developed a successful algorithm for solving equivalent reconstruction problems in optics (where $|\psi(q)|^2$ and $|\tilde{\psi}(p)|^2$ play the role of near-field and far-field intensity, for instance). A. Orlowski and H. Paul [205] applied a typical quantum-mechanical formalism to perform the Pauli reconstruction. Z. Bialynicka-Birula and I. Bialynicki-Birula [30] and J.A. Vaccaro and S.M. Barnett [275] solved the Pauli problem for photon number and quantum-optical phase instead of position and momentum. E. Feenberg [94] showed that the wave function can be inferred from the position probability distribution $|\psi(x, t)|^2$ and its temporal derivative $\partial |\psi(x, t)|^2 / \partial t$; see also the book [133] by E.C. Kemble. Note, however, that the extension of the proof [133] to three dimensions is wrong [99]. See also the interesting paper [287] by S. Weigert.

R.G. Newton and B.-L. Young [195] invented a recipe to measure the spin density matrix. Later W. Band and J.L. Park [12]–[15] developed a general procedure for solving the state-inference problem and gave explicit examples for spin 1/2, spin 1, and one-dimensional spinless systems. I.D. Ivanović [124] refined this method. Another paper by Ivanović [123] served as the mathematical basis for W.K. Wootters' work on the subject. See Refs. [297], [298]. It was further developed by W.K. Wootters and B.D. Fields [299]. U. Larsen [147] related it to the concept of complementary aspects. Other "early" ideas on state reconstruction were developed by A. Royer [238], [239] and M. Wilkens and P. Meystre [291].

The first practical demonstration of quantum tomography (optical homodyne tomography) by D.T. Smithey, M. Beck, M.G. Raymer, and A. Faridani [255] initiated a remarkable series of papers on state reconstruction. This field is still rapidly growing and diversifying. So it seems to be wise to refer the reader to the current literature instead of giving an incomplete list of references.

2

Quantum theory of light

2.1 The electromagnetic oscillator

Light shows both wave and particle aspects. It propagates in space and interferes with itself, it disperses in optical media such as prisms, and it displays polarization effects. All these properties are commonly regarded as wave features. On the other hand, when detected with sufficiently high precision, light appears as distinct detector clicks called photons. We may say as well that light behaves like moving particles that nevertheless follow the rules of wave interference. This strange picture has puzzled countless people during much of this century. Strictly speaking, the picture has not been explained yet, but rather it has been formulated more precisely in the quantum theory of light. According to this theory, the wave features of light are regarded as classical aspects (which does not necessarily mean that the particle aspects are entirely quantum). This book focuses on the quantum aspects of light. We will use the most primitive concept for the classical wave features but a sophisticated machinery for the quantum aspects. Our model is the electromagnetic oscillator. One complex vector function $u(x, t)$ called a *spatial–temporal mode* comprises all classical wave aspects including polarization. The simplest example of a spatial–temporal mode is a plane wave

$$u(x, t) = u_0 \exp[i(kx - \omega t)] \tag{2.1}$$

of polarization vector u_0, frequency ω, and wave vector k with $k^2 = \omega^2/c^2$. (As usual, c denotes the speed of light.) This mode defines a framework in space and time that may be excited by the quantum field "light." The mode function quantifies the strength of one excitation in space and time. Of course, the possibilities for setting the frame $u(x, t)$ are infinite as long as the spatial–temporal mode obeys the laws of classical waves, that is, Maxwell's equations. The choice of $u(x, t)$ is made by the observer. (We will study in Section 4.2.3 how this is accomplished in a particular type of experiment.) The observer singles out one mode, one quantum object, from the rest of the world. This object turns out to be a harmonic oscillator described by the annihilation operator $\hat{a}$. The operator $\hat{a}$ stands for the quantized amplitude with which the spatial–temporal mode can be excited. In classical optics it would be just a complex number α of magnitude $|\alpha|$ and phase arg α. The quantized amplitude $\hat{a}$ is neither predetermined nor given by the observer but depends on the state of the spatial–temporal mode. This state exists even if literally nothing is in the mode chosen by the observer. Then the light is just in the *vacuum state*. We will see later in this book that this "nothing" can indeed cause significant physical effects.

To make all these woolly words more precise and to cut a long story short, we postulate that the electric field strength $\hat{E}$ of the light field is given by

$$\hat{E} = u^*(x, t)\hat{a} + u(x, t)\hat{a}^\dagger \qquad (2.2)$$

and that the amplitude operator $\hat{a}$ is a bosonic annihilation operator, that is, $\hat{a}$ obeys the commutation relation

$$[\hat{a}, \hat{a}^\dagger] = 1. \qquad (2.3)$$

Throughout this book we set Planck's constant

$$\hbar = 1 \qquad (2.4)$$

for simplicity. (This can be always achieved by a proper rescaling of physical units.)

In the following we introduce the key elements of quantum-oscillator physics. The *photon-number* operator $\hat{n}$ accounts for the photons in the chosen spatial–temporal mode and is given by the counterpart of a classical modulus-squared amplitude

$$\hat{n} \equiv \hat{a}^\dagger \hat{a}. \qquad (2.5)$$

We introduce the *phase-shifting* operator

$$\hat{U}(\theta) \equiv \exp(-\mathrm{i}\theta\hat{n}). \qquad (2.6)$$

As the name suggests, the phase-shifting operator provides the amplitude $\hat{a}$ with a phase shift θ when acting on $\hat{a}$

$$\hat{U}^\dagger(\theta)\hat{a}\hat{U}(\theta) = \hat{a}\exp(-\mathrm{i}\theta). \qquad (2.7)$$

This property is easily seen by calculating the derivative of $\hat{U}^\dagger\hat{a}\hat{U}$ with respect to θ

$$\frac{d}{d\theta}\hat{U}^\dagger\hat{a}\hat{U} = \mathrm{i}[\hat{n}, \hat{U}^\dagger\hat{a}\hat{U}] = \hat{U}^\dagger\mathrm{i}[\hat{n}, \hat{a}]\hat{U}$$
$$= -\mathrm{i}\hat{U}^\dagger[\hat{a}, \hat{a}^\dagger]\hat{a}\hat{U} = -\mathrm{i}\hat{U}^\dagger\hat{a}\hat{U}. \qquad (2.8)$$

Because the right-hand side of Eq. (2.7) obeys the same differential equation with the initial operator $\hat{a}$ for $\theta = 0$, both sides must be equal indeed. There is another way of looking at formula (2.7). When the observer wishes to change the phase of the spatial–temporal mode, that is,

$$\hat{E} = u^*(x, t) \exp(-i\theta)\hat{a} + u(x, t) \exp(+i\theta)\hat{a}^\dagger, \tag{2.9}$$

this phase is picked up by the quantum amplitude $\hat{a}$. We may understand this to mean that the field state $\hat{\rho}$ has been altered by the observer to produce a new state

$$\hat{\rho}(\theta) = \hat{U}\hat{\rho}\hat{U}^\dagger \tag{2.10}$$

because any predictable quantity or expectation value depending on $\hat{a}$ $\exp(-i\theta) = \hat{U}^\dagger\hat{a}\hat{U}$ is reproduced when $\hat{\rho}$ is replaced by $\hat{\rho}(\theta)$ and $\hat{a}$ is not touched. In formulas,

$$\mathrm{tr}\{F[\hat{a}\exp(-i\theta)]\hat{\rho}\} = \mathrm{tr}\{F(\hat{U}^\dagger\hat{a}\hat{U})\hat{\rho}\}$$
$$= \mathrm{tr}\{F(\hat{a})\hat{U}\hat{\rho}\hat{U}^\dagger\}, \tag{2.11}$$

as is easily verified by expanding F in powers of $\hat{a}$. We note that this idea of replacing a change in the observables by a change of the state is no different from the transition from a Heisenberg to a Schrödinger picture in quantum mechanics.

Finally, we introduce a pair of operators, $\hat{q}$ and $\hat{p}$, called the *quadratures*. They appear as the "real" and the "imaginary" part, respectively, of the "complex" amplitude $\hat{a}$ multiplied by $2^{1/2}$:

$$\hat{q} = 2^{-1/2}(\hat{a}^\dagger + \hat{a}), \qquad \hat{p} = i2^{-1/2}(\hat{a}^\dagger - \hat{a}) \tag{2.12}$$

so that

$$\hat{a} = 2^{-1/2}(\hat{q} + i\hat{p}). \tag{2.13}$$

In optics $\hat{q}$ and $\hat{p}$ correspond to the in-phase and the out-of-phase component of the electric field amplitude of the spatial–temporal mode (with respect to a reference phase). It is easy to see from the basic bosonic commutation relation (2.3) that $\hat{q}$ and $\hat{p}$ are canonically conjugate observables,

$$[\hat{q}, \hat{p}] = i. \tag{2.14}$$

(Note that $\hbar = 1$.) The quadratures $\hat{q}$ and $\hat{p}$ can be regarded as the position and the momentum of the electromagnetic oscillator. Of course, they do not appear in real space but in the phase space spanned by the complex vibrational amplitude $\hat{a}$ of the electromagnetic oscillator, and they have nothing to do with the position and the momentum of a photon (concepts that are problematic in any case). Nevertheless, the canonical commutation relation (2.14) entitles us to treat $\hat{q}$ and $\hat{p}$ as perfect examples of position- and momentumlike quantities. We will see later in this book that this analogy is one of the key points why quantum optics allows some fundamental Gedanken experiments of quantum physics to be carried out – not literally but certainly in the spirit of their inventors. We note that phase shifting rotates the quadratures

$$\hat{q}_\theta \equiv \hat{U}^\dagger(\theta)\hat{q}\hat{U}(\theta) = \hat{q}\cos\theta + \hat{p}\sin\theta \tag{2.15}$$

$$\hat{p}_\theta \equiv \hat{U}^\dagger(\theta)\hat{p}\hat{U}(\theta) = -\hat{q}\sin\theta + \hat{p}\cos\theta, \tag{2.16}$$

as is easily verified using definition (2.12) and the phase-shifting property (2.7) of the annihilation operator $\hat{a}$. We see that we can go from a position representation to a momentum representation via a phase shift θ of $\pi/2$. Finally, we express the photon-number operator $\hat{n}$ in terms of the quadratures $\hat{q}$ and $\hat{p}$ and obtain, using the canonical commutation relation (2.14),

$$\hat{H} \equiv \hat{n} + \frac{1}{2} = \frac{\hat{q}^2}{2} + \frac{\hat{p}^2}{2}. \tag{2.17}$$

The right-hand side of this equation stands for the energy of a harmonic oscillator with unity mass and frequency, that is, the photon number plus $1/2$ gives the energy of the electromagnetic oscillator. The additional $1/2$ is called the *vacuum energy* for a reason explained in Section 2.2.2.

2.2 Single-mode states

In this section we introduce several states of the electromagnetic oscillator that have a number of useful applications (in a purely mathematical or in a truly physical sense). We begin with the quadrature states, then turn to the Fock states, and consider finally coherent states as the most important realistic states of light. All states are introduced as eigenstates of prominent observables such as the quadratures, the photon number, and the annihilation operator.

2.2.1 Quadrature states

Let us call the eigenstates $|q\rangle$ and $|p\rangle$ of the quadratures $\hat{q}$ and $\hat{p}$ *quadrature states,* satisfying

$$\hat{q}|q\rangle = q|q\rangle, \qquad \hat{p}|p\rangle = p|p\rangle. \tag{2.18}$$

Because the quadratures obey the canonical commutation relation (2.14) their spectrum must be unbounded and continuous [58], as we would expect for position and momentum (see also Section 6.3). They are orthogonal

$$\langle q\,|q'\rangle = \delta(q-q'), \qquad \langle p\,|p'\rangle = \delta(p-p') \tag{2.19}$$

and complete

$$\int_{-\infty}^{+\infty} |q\rangle\langle q|\, dq = \int_{-\infty}^{+\infty} |p\rangle\langle p|\, dp = 1. \tag{2.20}$$

As is well known, position and momentum states are mutually related to each other by Fourier transformation

$$|q\rangle = \frac{1}{\sqrt{2\pi}} \int_{-\infty}^{+\infty} \exp(-iqp)|p\rangle\, dp \tag{2.21}$$

$$|p\rangle = \frac{1}{\sqrt{2\pi}} \int_{-\infty}^{+\infty} \exp(+iqp)|q\rangle\, dq. \tag{2.22}$$

However, the quadrature states are not truly normalizable, and so they cannot be generated experimentally (at least in a strict sense). Nevertheless, they will appear in many mathematical tricks. For instance, they are needed to introduce the *quadrature wave functions*

$$\psi(q) = \langle q \mid \psi\rangle, \qquad \tilde{\psi}(p) = \langle p \mid \psi\rangle. \tag{2.23}$$

In contrast to the quadrature states, the quadrature wave functions have a physical meaning. Their moduli squared account for the quadrature probability distributions $|\psi(q)|^2$ and $|\tilde{\psi}(p)|^2$ of the pure state $|\psi\rangle$, which can be precisely measured using homodyne detection, as will be considered in detail in Section 4.2.

2.2.2 Fock states

Let us introduce *Fock states*, or $|n\rangle$, as the eigenstates of the photon-number operator $\hat{n}$

$$\hat{n}|n\rangle = n|n\rangle. \tag{2.24}$$

Fock states are named after the Russian physicist V.A. Fock and are widely used in quantum field theory. As eigenstates of the number operator $\hat{n}$, Fock states have a perfectly fixed photon number. They possess appealing physical properties but are difficult to generate with present technology; see for instance Refs. [117] and [134] and the references cited therein.

Let us study the Fock states in some detail. First, we see that if $|n\rangle$ is an eigenstate of $\hat{n}$, then $\hat{a}|n\rangle$ must be an eigenstate as well, with the eigenvalue $n - 1$. In fact,

$$\hat{n}\hat{a}|n\rangle = \hat{a}^\dagger\hat{a}^2|n\rangle = \left(\hat{a}\hat{a}^\dagger\hat{a} - \hat{a}\right)|n\rangle = (n-1)\hat{a}|n\rangle. \tag{2.25}$$

In a similar way we easily show that $\hat{a}^\dagger|n\rangle$ is an eigenstate of $\hat{n}$ with the eigenvalue $n + 1$. So we derive the fundamental relations

$$\hat{a}|n\rangle = \sqrt{n}|n - 1\rangle, \tag{2.26}$$

$$\hat{a}^\dagger|n\rangle = \sqrt{n + 1}|n + 1\rangle. \tag{2.27}$$

The prefactors have been obtained using the fact that $\langle n|\hat{a}^\dagger\hat{a}|n\rangle$ must equal the eigenvalue n. Because of these relations, $\hat{a}$ is called the *annihilation operator* (it takes one photon out of $|n-1\rangle$) and $\hat{a}^\dagger$ is called the *creation operator*. The annihilation operator or the creation operator lowers or raises the photon number in integer steps. What would happen if we had a Fock state with noninteger eigenvalue n? A sufficiently large number of lowerings would certainly produce a Fock state with a photon number less than $-1/2$. On the other hand, we know from the relation (2.17) of $\hat{n}$ to the energy $\hat{H}$ that the average

$$\langle\hat{n}\rangle = \left\langle\frac{\hat{q}^2}{2}\right\rangle + \left\langle\frac{\hat{p}^2}{2}\right\rangle - \frac{1}{2} \geq -\frac{1}{2}. \qquad (2.28)$$

This bound leads to a contradiction, because for eigenstates of $\hat{n}$ the average $\langle\hat{n}\rangle$ should equal the eigenvalue n. Consequently, no fractional photons exist, at least if the photon number is fixed precisely, that is, for photon-number eigenstates.

What happens if the photon number is integer – if we reach zero after lowering n in integer steps? Two options satisfy

$$\hat{a}^\dagger\hat{a}|0\rangle = 0. \qquad (2.29)$$

One is to require that

$$\hat{a}|0\rangle = 0. \qquad (2.30)$$

The other that

$$\hat{a}|0\rangle \neq 0, \quad \text{but} \quad \hat{a}^\dagger(\hat{a}|0\rangle) = 0. \qquad (2.31)$$

Let us study the first option (2.30) first. Using the quadrature decomposition (2.13) of the annihilation operator and Schrödinger's famous formula $\hat{p} = -i\partial/\partial q$ in the q-representation, we obtain a differential equation for the wave function $\psi_0(q)$ of the state $|0\rangle$

$$\hat{a}\psi_0(q) = \frac{1}{\sqrt{2}}\left(q + \frac{\partial}{\partial q}\right)\psi_0(q) = 0. \qquad (2.32)$$

The solution of this equation is

$$\psi_0(q) = \pi^{-1/4}\exp\left(-\frac{q^2}{2}\right) \qquad (2.33)$$

(normalized to yield $\int_{-\infty}^{+\infty}|\psi_0(q)|^2\,dq = 1$). In the momentum representation we obtain the same formula for $\tilde{\psi}_0(p)$

$$\tilde{\psi}_0(p) = \pi^{-1/4}\exp\left(-\frac{p^2}{2}\right). \qquad (2.34)$$

In this way we have shown that a well-behaved state with precisely zero photons called the *vacuum state* exists. So even if the spatial–temporal mode is

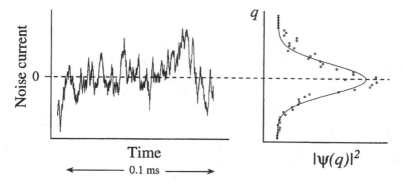

Fig. 2.1. Measurement of the vacuum noise. The position quadrature of an empty field was measured using balanced homodyne detection (see Section 4.2). Although the measurement time (0.1 milliseconds) is rather short, so that the number of samples is relatively low, the histogram of the noise current (dots) is approximately Gaussian and already follows the theoretical expectation (solid curve). [Courtesy of G. Breitenbach, University of Constance.]

completely empty, a physically meaningful state that might cause physical effects is still associated with this "emptiness."* Figure 2.1 shows a plot of the quadrature probability distribution $|\psi_0(q)|^2$ for a vacuum that has been measured using homodyne detection. (For an analysis of homodyne detection, see Section 4.2.) This curve illustrates beautifully that even in a complete vacuum the quadratures are still restlessly fluctuating. (This is the zero-point motion.) Of course, they must fluctuate; if both position and momentum quadratures were fixed, Heisenberg's uncertainty principle would be violated. The fluctuation energy of the vacuum state gives rise to the vacuum term $1/2$ in the energy (2.17) of the electromagnetic oscillator.

Excited states are solutions of the relation (2.27) for an initial vacuum

$$|n\rangle = \frac{\hat{a}^{\dagger n}}{\sqrt{n!}}|0\rangle. \tag{2.35}$$

We obtain a formula for their wave functions by expressing the relation (2.27) for $n = m - 1$ in the Schrödinger representation

$$\hat{a}^{\dagger}\psi_{m-1}(q) = \frac{1}{\sqrt{2}}\left(q - \frac{\partial}{\partial q}\right)\psi_{m-1}(q) = \sqrt{m}\psi_m(q). \tag{2.36}$$

This formula is satisfied by

$$\psi_n(q) = \frac{H_n(q)}{\sqrt{2^n n! \sqrt{\pi}}} \exp\left(-\frac{q^2}{2}\right). \tag{2.37}$$

*Throughout this book we always mean by "vacuum" simply "no light" and not an evacuated system.

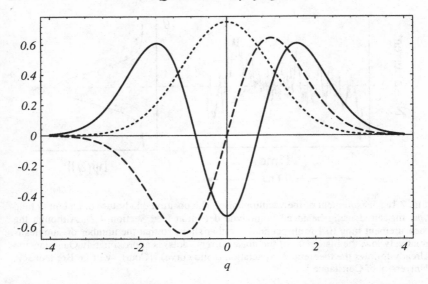

Fig. 2.2. Plot of the quadrature wave functions for some Fock states. Dotted line: vacuum ($n = 0$), dashed line: first excited state ($n = 1$), and solid line: second excited state ($n = 2$). The wave functions are even, $\psi_n(-q) = \psi_n(q)$, for even numbers n and odd, $\psi_n(-q) = -\psi_n(q)$, for odd numbers. They oscillate in the classically allowed region between the turning points of a classical harmonic oscillator with energy $n + 1/2$. Outside this region, that is, in the classically forbidden zone, the wave functions decay exponentially.

Here the H_n denote the Hermite polynomials, and we have used relation 10.13(14) of Ref. [89], Vol. II. Because we know that the Fock wave function for $n = 0$ is the vacuum wave function ψ_0 given by Eq. (2.33), we have found the ψ_n uniquely. Figure 2.2 shows plots of some Fock wave functions. They appear as standing Schrödinger waves for quadrature values ranging between the Bohr–Sommerfeld bands $-(2n+1)^{1/2}$ and $(2n+1)^{1/2}$. This behavior can be verified using the semiclassical theory for energy eigenstates described in Appendix 1. Consequently, the quadrature distributions, or squared wave functions, are broad. They illustrate that because the Fock states are particlelike, they have noisy quadrature amplitudes and exhibit few features of a classical, stable wave.

Let us return to the second possibility (2.31) for a vacuum state with a wave function $\varphi_0(q)$. It means that the function

$$\varphi_{-1}(q) \equiv \hat{a}\varphi_0(q) = \frac{1}{\sqrt{2}}\left(q + \frac{\partial}{\partial q}\right)\varphi_0(q) \qquad (2.38)$$

satisfies

$$\hat{a}^\dagger\varphi_{-1}(q) = \frac{1}{\sqrt{2}}\left(q - \frac{\partial}{\partial q}\right)\varphi_{-1}(q) = 0. \qquad (2.39)$$

However, the solution of this equation

$$\varphi_{-1}(q) = c \exp\left(+\frac{q^2}{2}\right) \tag{2.40}$$

is not normalizable. Hence $\varphi_0(q)$, or the inhomogeneous solution of the differential equation (2.38), is not normalizable as well. It is called the *irregular wave function* of the vacuum state and is given by the expression

$$\varphi_0(q) = c\sqrt{\frac{\pi}{2}} \exp\left(-\frac{q^2}{2}\right) \text{erfi}(q) \tag{2.41}$$

with erfi being the imaginary error function

$$\text{erfi}(z) = \frac{2}{\sqrt{\pi}} \int_0^z \exp(t^2)\, dt. \tag{2.42}$$

Because the irregular vacuum is not normalizable, it must be rejected as a physically meaningful state. However, we will encounter later a useful mathematical application of irregular wave functions.

We have shown that a unique vacuum state exists for the electromagnetic oscillator and that all Fock states are given as excitations (2.35) of the vacuum. (Because the Schrödinger equation is of second order there are only two fundamental solutions – regular and irregular wave functions. The irregular ones are discarded as physical states.) So the Fock states must be complete,

$$\sum_{n=0}^{\infty} |n\rangle\langle n| = 1, \tag{2.43}$$

that is, they span the whole Hilbert space of the electromagnetic oscillator. Additionally, Fock states are orthonormal,

$$\langle n \mid n'\rangle = \delta_{nn'}, \tag{2.44}$$

because they are eigenstates of the Hermitian operator $\hat{n}$. The Fock states form the most convenient and most frequently used orthonormal Hilbert-space basis in quantum optics, called the *Fock basis*.

2.2.3 Coherent states

We introduce *coherent states* as the eigenstates of the annihilation operator $\hat{a}$

$$\hat{a}|\alpha\rangle = \alpha|\alpha\rangle. \tag{2.45}$$

Coherent states are also named *Glauber states* after the American physicist R.J. Glauber. They are called coherent states because light fields in these states are perfectly coherent in the sense of Ref. [187]. High-quality lasers generate such fields. As eigenstates of the annihilation operator $\hat{a}$, the coherent states have well-defined amplitudes $|\alpha|$ and phases arg α. (Because the annihilation operator $\hat{a}$ is not Hermitian, the eigenvalues of $\hat{a}$ are complex. They correspond to the complex wave amplitudes in classical optics.) Coherent states come as close as quantum mechanics allows to wavelike states of the electromagnetic oscillator. Because the wave aspects of light are commonly regarded as classical, coherent states are often called classical states. Furthermore, fields in statistical mixtures of coherent states (such as thermal fields) are classical as well, whereas any state that cannot be understood as an ensemble of coherent states is called *nonclassical*. The experimental generation and application of nonclassical light fields is one of the top issues of modern quantum optics. Despite much progress made, producing nonclassical states is still extremely challenging because they are easily destroyed (reduced to classical) by any kind of losses.

Let us return to coherent states and study their properties. First we note that vacuum is a coherent state as well, because it satisfies Eq. (2.45) for $\alpha = 0$, that is, vacuum is a zero-amplitude coherent state. Without much mathematical effort we see directly from the definition (2.45) that the mean energy of a coherent state is simply

$$\langle \hat{H} \rangle = \langle \alpha | \hat{a}^\dagger \hat{a} + \frac{1}{2} | \alpha \rangle = |\alpha|^2 + \frac{1}{2}, \qquad (2.46)$$

or the sum of the classical wave intensity $|\alpha|^2$ and the vacuum energy $1/2$. We also see easily from the definition (2.45) that a phase shift by the angle θ simply shifts the phase arg α of the coherent-state amplitude

$$\hat{U}(\theta)|\alpha\rangle = |\alpha \exp(-i\theta)\rangle. \qquad (2.47)$$

This result is what we would expect for wavelike states.

To study coherent states more carefully, we introduce the *displacement operator*

$$\hat{D}(\alpha) = \exp(\alpha \hat{a}^\dagger - \alpha^* \hat{a}). \qquad (2.48)$$

Because $i(\alpha \hat{a}^\dagger - \alpha^* \hat{a})$ is Hermitian, $\hat{D}$ must be unitary. The displacement operator displaces the amplitude $\hat{a}$ by the complex number α

$$\hat{D}^\dagger(\alpha)\hat{a}\hat{D}(\alpha) = \hat{a} + \alpha. \qquad (2.49)$$

To prove this statement we imagine the displacement α as being decomposed into infinitesimal steps $\delta\alpha$ and we obtain in first order of $\delta\alpha$

$$\hat{D}^\dagger(\delta\alpha)\hat{a}\hat{D}(\delta\alpha) = \hat{a} + \left[\hat{a}, \hat{a}^\dagger\delta\alpha - \hat{a}\delta\alpha^*\right] = \hat{a} + \delta\alpha. \qquad (2.50)$$

Because the total displacement operator $\hat{D}(\alpha)$ with $\alpha = \sum \delta\alpha$ is the product $\Pi \hat{D}(\delta\alpha)$ of the infinitesimal displacements $\hat{D}(\delta\alpha)$, we can apply the infinitesimal steps (2.50) as often as we need to show that $\hat{D}^\dagger(\alpha)\hat{a}\hat{D}(\alpha)$ equals indeed $\hat{a} + \sum \delta\alpha$, which proves Eq. (2.49). (Readers who are familiar with Lie groups notice easily that we have used the properties of the Lie algebra to study the group elements. Quite generally, coherent states are intimately linked to Lie groups; see Ref. [216]). What has the displacement operator to do with coherent states? Let us apply a "negative" displacement to $|\alpha\rangle$. We see from the basic property (2.49) of the displacement operator that

$$\hat{a}\hat{D}(-\alpha)|\alpha\rangle = \hat{D}(-\alpha)\hat{D}^\dagger(-\alpha)\hat{a}\hat{D}(-\alpha)|\alpha\rangle$$
$$= \hat{D}(-\alpha)(\hat{a} - \alpha)|\alpha\rangle, \qquad (2.51)$$

which must equal zero because of the definition (2.45) of coherent states. This implies that $\hat{D}(-\alpha)|\alpha\rangle$ is *the* vacuum state $|0\rangle$. Consequently, coherent states $|\alpha\rangle$ are *displaced vacuums*

$$|\alpha\rangle = \hat{D}(\alpha)|0\rangle. \qquad (2.52)$$

Of course, this does not mean that coherent states are physically similar to vacuum states. They have only some quantum-noise properties in common. (Displacing the vacuum may appear as rather inappropriate description of generating high-quality laser light.)

To study the relation between coherent states and the vacuum in more detail, we calculate the quadrature wave functions $\psi_\alpha(q)$ and $\tilde{\psi}_\alpha(p)$. We decompose the complex amplitude α into real and imaginary parts

$$\alpha = 2^{-1/2}(q_0 + ip_0) \qquad (2.53)$$

and represent the displacement operator in terms of the quadratures $\hat{q}$ and $\hat{p}$

$$\hat{D} = \exp(ip_0\hat{q} - iq_0\hat{p}). \qquad (2.54)$$

We take advantage of the *Baker–Hausdorff formula*

$$\exp(\hat{F} + \hat{G}) = \exp\left(-\frac{1}{2}[\hat{F}, \hat{G}]\right)\exp(\hat{F})\exp(\hat{G})$$
$$= \exp\left(+\frac{1}{2}[\hat{F}, \hat{G}]\right)\exp(\hat{G})\exp(\hat{F}) \qquad (2.55)$$

for any two operators $\hat{F}$ and $\hat{G}$ such that the commutator $[\hat{F}, \hat{G}]$ commutes with both of them. This fundamental operator relation is proven for instance in Ref. [100]. Here we use the Baker–Hausdorff formula to split $\hat{D}$ into three parts

$$\hat{D}(\alpha) = \exp\left(-\frac{ip_0q_0}{2}\right)\exp(ip_0\hat{q})\exp(-iq_0\hat{p}) \qquad (2.56)$$

$$= \exp\left(+\frac{ip_0q_0}{2}\right)\exp(-iq_0\hat{p})\exp(ip_0\hat{q}). \qquad (2.57)$$

In the position representation the momentum operator $\hat{p}$ equals $-i\partial/\partial q$ and the exponential $\exp(-q_0\partial/\partial q)$ is a translation operator

$$\exp\left(-q_0\frac{\partial}{\partial q}\right)\psi(q) = \psi(q - q_0), \tag{2.58}$$

as it is easily verified by differentiating both sides with respect to q_0. In this way we realize that the displacement operator acts in three steps on position wave functions: First it displaces the wave function; then it multiplies it with $\exp(i p_0 \hat{q})$, that is, with $\exp(i p_0 q)$ in the position representation; and finally, the displacement operator attaches the phase factor $\exp(-i p_0 q_0/2)$ to the wave function. Because coherent states are displaced vacuums the position wave function is simply

$$\psi_\alpha(q) = \psi_0(q - q_0)\exp\left(+i p_0 q - \frac{i p_0 q_0}{2}\right) \tag{2.59}$$

$$= \pi^{-1/4}\exp\left[-\frac{(q - q_0)^2}{2} + i p_0 q - \frac{i p_0 q_0}{2}\right], \tag{2.60}$$

with ψ_0 being the vacuum wave function. In a similar way we obtain the momentum wave function

$$\tilde{\psi}_\alpha(p) = \pi^{-1/4}\exp\left[-\frac{(p - p_0)^2}{2} - i q_0 p + \frac{i p_0 q_0}{2}\right]. \tag{2.61}$$

The Eqs. (2.60) and (2.61) show that the quadrature probability distributions $|\psi_\alpha(q)|^2$ and $|\tilde{\psi}_\alpha(p)|^2$ of coherent states are Gaussian with the same width as the Gaussian curve for vacuum. They are shifted only by the real amplitudes q_0 and p_0, and we would obtain a completely similar picture for the coherent states as for the vacuum in Fig. 2.1 (see Fig. 3.2). In this sense coherent states are similar to the vacuum. Only the vacuum fluctuations contaminate the quadrature amplitudes, illustrating that coherent states are wavelike – they have just as much quadrature noise as is unavoidable. This reason is why high-quality laser light is a wonderful tool for experimentation.

So much for the wave features of coherent states; let us now study the particle aspects. For this purpose we seek the Fock representation of $\hat{D}(\alpha)|0\rangle$. We express $\hat{D}$ in terms of $\hat{a}$ and $\hat{a}^\dagger$, Eq. (2.48), and use the Baker–Hausdorff formula (2.55) again to split the displacement operator into three parts

$$\hat{D} = \exp\left(-\frac{1}{2}|\alpha|^2\right)\exp\left(\alpha\hat{a}^\dagger\right)\exp(-\alpha^*\hat{a}). \tag{2.62}$$

Because the annihilation operator annihilates the vacuum, all powers of $\hat{a}$ contained in the Taylor series expansion of the exponential $\exp(-\alpha^*\hat{a})$ give zero when applied to $|0\rangle$, with the only exception being the zeroth order term 1. So the exponential $\exp(-\alpha^*\hat{a})$ does not affect the vacuum state at all. We expand the other exponential $\exp(\alpha\hat{a}^\dagger)$ in the splitting (2.62) in the Taylor series and

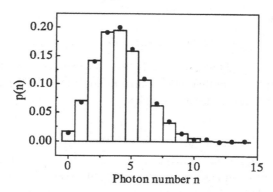

Fig. 2.3. Photon-number distribution of light in a coherent state. The distribution was obtained from experimental homodyne data via the method described in Section 5.2. We see that the reconstructed histogram (dots) is approximately Poissonian (bar chart). [Courtesy of G. Breitenbach, University of Constance.]

use formula (2.35) for the Fock states $|n\rangle$ to obtain

$$|\alpha\rangle = \exp\left(-\frac{1}{2}|\alpha|^2\right) \sum_{n=0}^{\infty} \frac{\alpha^n}{\sqrt{n!}} |n\rangle. \tag{2.63}$$

The Fock representation (2.63) shows that a coherent state has Poissonian photon statistics

$$p_n = |\langle n \mid \alpha \rangle|^2 = \frac{|\alpha|^{2n}}{n!} \exp(-|\alpha|^2). \tag{2.64}$$

Counting the photons of a coherent state means making repeated measurements on a statistical ensemble of identically prepared fields. Each time, n photons are obtained with Poissonian probability p_n, and on average we get as many photons as quantified by the intensity $|\alpha|^2$. Classical particles obey the same statistical law when they are taken at random from a pool with an average of $|\alpha|^2$ each time. We may say that when the photons of a coherent state are counted they behave like randomly distributed classical particles. This classical randomness seems not too surprising because coherent states are wavelike.

Let us finally derive some formal properties of coherent states that turn out to be quite useful. First, we note that coherent states are not exactly orthogonal to each other because they are not eigenstates of a Hermitian operator. Instead, they are approximately orthogonal when their amplitudes differ sufficiently. In fact, we obtain from the Fock representation (2.63)

$$\langle \alpha' \mid \alpha \rangle = \exp\left(-\frac{|\alpha|^2}{2} - \frac{|\alpha'|^2}{2}\right) \sum_{n=0}^{\infty} \frac{(\alpha'^*\alpha)^n}{n!}$$

$$= \exp\left(-\frac{|\alpha|^2}{2} - \frac{|\alpha'|^2}{2} + \alpha'^*\alpha\right) \tag{2.65}$$

and consequently

$$|\langle \alpha' | \alpha \rangle|^2 = \exp(-|\alpha - \alpha'|^2). \tag{2.66}$$

The Gaussian in Eq. (2.66) goes rapidly to zero when the amplitudes α and α' differ significantly more than the quadrature-noise level of the vacuum. We also note that coherent states are complete,

$$\int_{-\infty}^{+\infty} \int_{-\infty}^{+\infty} |\alpha\rangle\langle\alpha| \frac{dq_0\, dp_0}{2\pi} = 1. \tag{2.67}$$

That is, we may express physical quantities in a coherent-state basis. (Coherent states are even overcomplete because fewer than all of them form a basis already. See Refs. [9] and [50]. This property is, by the way, a side effect of their lack of strict orthogonality.) The proof of the completeness relation (2.67) is a matter of substituting the Fock representation for $|\alpha\rangle$ and performing the necessary integrations using polar coordinates in the complex plane.

2.3 Uncertainty and squeezing

In the preceding section we introduced Fock states and coherent states as some physically meaningful states of the electromagnetic oscillator. We have seen that the quadrature amplitudes of these states fluctuate according to certain probability distributions. Coherent states are distinguished for having only as much statistical uncertainty in their quadrature amplitudes as the vacuum. Is this the quantum-mechanical optimum, or, to put it differently, what are the *minimum-uncertainty states*? Pauli settled this question in a brilliant few-line proof published in his *Handbuch der Physik* article [212]. Let us first denote the average complex amplitude of possible candidate states $|\psi\rangle$ by α

$$\langle \psi | \hat{a} | \psi \rangle = \alpha = 2^{-1/2}(q_0 + ip_0). \tag{2.68}$$

We remove the amplitude α from consideration by applying the displacement operator to obtain a new state, $|\varphi\rangle$,

$$|\varphi\rangle = \hat{D}(-\alpha)|\psi\rangle \tag{2.69}$$

that contains the same amount of quadrature noise as the state $|\psi\rangle$. The variances of q and p are given by

$$\Delta^2 q = \langle \psi | (\hat{q} - q_0)^2 | \psi \rangle = \langle \varphi | \hat{q}^2 | \varphi \rangle \tag{2.70}$$

and

$$\Delta^2 p = \langle \psi | (\hat{p} - p_0)^2 | \psi \rangle = \langle \varphi | \hat{p}^2 | \varphi \rangle. \tag{2.71}$$

Now we use Pauli's argument [212] and state that for the position wave function $\varphi(q)$ of $|\varphi\rangle$ the quantity

$$\delta \equiv \left| \frac{q}{2\Delta^2 q} \varphi + \frac{\partial \varphi}{\partial q} \right|^2 \tag{2.72}$$

must necessarily be greater than or at least equal to zero. On the other hand,

$$\delta = \frac{1}{4} \left(\frac{q}{\Delta^2 q} \right)^2 \varphi^* \varphi + \frac{q}{2\Delta^2 q} \left(\varphi \frac{\partial \varphi^*}{\partial q} + \varphi^* \frac{\partial \varphi}{\partial q} \right) + \frac{\partial \varphi^*}{\partial q} \frac{\partial \varphi}{\partial q}$$

$$= \frac{1}{4} \left(\frac{q}{\Delta^2 q} \right)^2 \varphi^* \varphi + \frac{1}{2} \frac{\partial}{\partial q} \left(\frac{q}{\Delta^2 q} \varphi^* \varphi \right) - \frac{1}{2} \frac{\varphi^* \varphi}{\Delta^2 q} + \frac{\partial \varphi^*}{\partial q} \frac{\partial \varphi}{\partial q}$$

$$= \frac{1}{4(\Delta^2 q)^2} (q^2 - 2\Delta^2 q) \varphi^* \varphi + \frac{\partial \varphi^*}{\partial q} \frac{\partial \varphi}{\partial q} + \frac{1}{2} \frac{\partial}{\partial q} \left(\frac{q}{\Delta^2 q} \varphi^* \varphi \right). \tag{2.73}$$

We integrate the last line, drop the total differential $\partial(\varphi^* \varphi q / \Delta^2 q)/\partial q$ and use $\hat{p} = -i\partial/\partial q$ for the $(\partial \varphi^*/\partial q) \cdot (\partial \varphi/\partial q)$ term to see that

$$\int_{-\infty}^{+\infty} \delta \, dq = -\frac{1}{4\Delta^2 q} + \Delta^2 p \geq 0 \tag{2.74}$$

which implies for $\Delta q = \sqrt{\Delta^2 q}$ and $\Delta p = \sqrt{\Delta^2 p}$

$$\Delta q \, \Delta p \geq \frac{1}{2}. \tag{2.75}$$

This formula is nothing else than Heisenberg's uncertainty relation (with $\hbar$ being scaled to unity). Because Pauli used the integration of δ in his derivation, he produced essentially a local version of Heisenberg's uncertainty principle. The great advantage of this mathematical trick is that it reveals the minimum-uncertainty states in the twinkling of an eye. In fact, because $\delta \geq 0$, the equality sign in relation (2.75) holds only if

$$\frac{1}{2} \frac{q}{\Delta^2 q} \varphi + \frac{\partial \varphi}{\partial q} = 0. \tag{2.76}$$

The normalized solution of this differential equation is

$$\varphi(q) = \left(\pi 2\Delta^2 q \right)^{-1/4} \exp \left[-\frac{q^2}{4\Delta^2 q} \right]. \tag{2.77}$$

So apart from a displacement, the minimum-uncertainty states have Gaussian wave functions like coherent states. However, the variance $\Delta^2 q$ should not necessarily equal $1/2$, as is the case for coherent states. In other words, both variances $\Delta^2 q$ and $\Delta^2 p$ are not required to be equal to minimize Heisenberg's uncertainty relation (2.75). The statistical uncertainty of the position quadrature q may be *squeezed* below the vacuum level $1/2$ at the cost, however, of enhancing the uncertainty in the canonically conjugate quadrature p and vice versa.

Let us study this squeezing effect more carefully. We parameterize the deviation of the variances from their vacuum values by a real number ζ called the *squeezing parameter*

$$\Delta^2 q = \frac{1}{2} e^{-2\zeta}, \qquad \Delta^2 p = \frac{1}{2} e^{+2\zeta}. \tag{2.78}$$

Obviously, the product of Δq and Δp equals the minimal value $1/2$. How can we squeeze the vacuum? Mathematically, we could just scale the position wave function ψ_0 for the vacuum

$$\varphi(q) = e^{\zeta/2} \psi_0(e^{\zeta} q). \tag{2.79}$$

The prefactor $e^{\zeta/2}$ in this scaling serves for maintaining the normalization of $\varphi(q)$. The momentum wave function $\tilde{\varphi}(p)$ is the Fourier-transformed position wave function. Consequently,

$$\tilde{\varphi}(p) = e^{-\zeta/2} \tilde{\psi}_0(e^{-\zeta} p). \tag{2.80}$$

This formula implies that the momentum wave function is antisqueezed when the position wave function is squeezed and vice versa. We differentiate φ in Eq. (2.79) with respect to the squeezing parameter ζ and obtain

$$\frac{\partial \varphi}{\partial \zeta} = \frac{1}{2}\left(q \frac{\partial}{\partial q} + \frac{\partial}{\partial q} q \right) \varphi = \frac{1}{2}(\mathrm{i}\hat{q}\hat{p} + \mathrm{i}\hat{p}\hat{q}) \varphi. \tag{2.81}$$

Since $\mathrm{i}\hat{q}\hat{p} + \mathrm{i}\hat{p}\hat{q}$ equals $\hat{a}^2 - \hat{a}^{\dagger 2}$ we can express the formal solution of this differential equation in terms of the unitary *squeezing operator*

$$\hat{S} \equiv \exp\left[\frac{\zeta}{2}\left(\hat{a}^2 - \hat{a}^{\dagger 2}\right)\right] \tag{2.82}$$

and obtain for the *squeezed-vacuum state*

$$|\varphi\rangle = \hat{S}(\zeta)|0\rangle. \tag{2.83}$$

According to Pauli's proof, all minimum-uncertainty states are displaced Gaussian states, that is, they have displaced rescaled vacuum wave functions. Consequently, all minimum-uncertainty states are *displaced squeezed vacuums*

$$|\psi\rangle = \hat{D}(\alpha)\hat{S}(\zeta)|0\rangle \tag{2.84}$$

having a position wave function of

$$\psi(q) = \pi^{-1/4} e^{\zeta/2} \exp\left[-e^{2\zeta} \frac{(q - q_0)^2}{2} + \mathrm{i}pq - \frac{\mathrm{i}p_0 q_0}{2}\right]. \tag{2.85}$$

In this way we have found not only a convenient mathematical notation for the squeezed states but also one possible physical process for generating squeezed light experimentally. We interpret simply the squeezing operator $\hat{S}(\zeta)$ as an evolution operator describing the result of the interaction

$$\hat{H}_{int} \propto \left(\hat{a}^2 - \hat{a}^{\dagger 2}\right). \tag{2.86}$$

The squeezing parameter ζ contains the product of the coupling strength and the interaction time. Processes described by nonlinear Hamiltonians such as $\hat{H}_{int}$ in Eq. (2.86) belong to the branch of *nonlinear optics*. In particular, the squeezing interaction (2.86) is realized by the degenerate parametric amplification of the spatial–temporal mode. A crystal such as KTP (potassium titanyl phosphate) is pumped by another laser beam at twice the frequency of the spatial–temporal mode of interest. The pump amplifies the signal parametrically much as a swing is amplified by changing the effective length at twice the frequency of the swing. A classical swing relies on tiny initial fluctuations that are in-phase with respect to the parametric pump. A quantum swing like the degenerate parametric amplifier experiences at least the vacuum fluctuations from the very beginning. Vacuum fluctuations that are in-phase with respect to the pump are amplified, whereas out-of-phase fluctuations get deamplified or, in other words, squeezed.

A squeezed vacuum requires a pump for generation, and, hence, when produced it carries energy. To quantify the amount of squeezing energy we note that the squeezing operator changes the quadratures

$$\hat{S}^{\dagger}(\zeta)\hat{q}\hat{S}(\zeta) = \hat{q}e^{-\zeta} \tag{2.87}$$

$$\hat{S}^{\dagger}(\zeta)\hat{p}\hat{S}(\zeta) = \hat{p}e^{+\zeta} \tag{2.88}$$

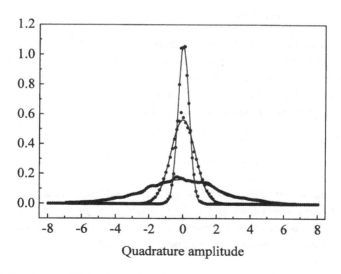

Fig. 2.4. Quadrature distributions of a squeezed vacuum [41]. Dots: measured values, lines: theoretical predictions. The squeezed and the antisqueezed components are depicted in comparison with the quadrature distribution of a vacuum (central plot).

because it scales the eigenfunctions of $\hat{q}$ and $\hat{p}$ accordingly. Substituting for $\hat{a}$ the quadrature decomposition (2.13) we see that

$$\hat{S}^{\dagger}(\zeta)\hat{a}\hat{S}(\zeta) = \hat{a}\cosh\zeta - \hat{a}^{\dagger}\sinh\zeta. \tag{2.89}$$

We use this formula to express the energy (2.17) of a squeezed state (2.84) and obtain

$$\langle\psi|\hat{H}|\psi\rangle = |\alpha|^2 + \frac{1}{2} + \sinh^2\zeta. \tag{2.90}$$

Three terms contribute to the energy. The first accounts for the coherent energy given by the modulus squared of the coherent amplitude α, the second is the vacuum energy $1/2$, and the last quantifies the fluctuation energy of squeezed states. Originally, the contribution to this squeezing energy comes from the pump used to generate the squeezed light. It is stored in the enhanced fluctuations of the antisqueezed component. Because both the squeezed and the antisqueezed quadratures contribute to $\hat{H} = (\hat{q}^2 + \hat{p}^2)/2$, even a squeezed vacuum carries energy.

Let us calculate the energy distribution, that is, the photon-number statistics of a squeezed vacuum

$$p_n = |\langle n|\hat{S}(\zeta)|0\rangle|^2. \tag{2.91}$$

We express the scalar product in the position representation

$$\langle n|\hat{S}(\zeta)|0\rangle = \int_{-\infty}^{+\infty} \psi_n(q)e^{\zeta/2}\psi_0(e^{\zeta}q)\,dq. \tag{2.92}$$

A squeezed vacuum as well as the vacuum state is perfectly symmetric if we flip the sign of the quadrature amplitude q. [It has an even wave function $\psi(q) = \psi(-q)$.] The wave functions $\psi_n(q)$ for the Fock states are even for even photon numbers and odd if n is odd. Consequently, the integral (2.92) vanishes for odd photon numbers and we obtain

$$p_{2m+1} = 0 \quad (m = 0, 1, 2, \ldots). \tag{2.93}$$

A squeezed vacuum contains only photon pairs. We may see this fact as a simple consequence of the mirror symmetry of squeezing. A deeper physical reason for this remarkable property is that a squeezed vacuum may be generated in a parametric process described by the quadratic Hamiltonian (2.86). Loosely speaking, photons are created in pairs: each pump photon is converted into two signal photons of half the pump frequency. The probability for finding a photon pair is

$$p_{2m} = \binom{2m}{m}\frac{1}{\cosh\zeta}\left(\frac{1}{2}\tanh\zeta\right)^{2m} \quad (m = 0, 1, 2, \ldots). \tag{2.94}$$

Here Eq. 2.20.3.3. of Ref. [225], Vol. II, has been used to perform the necessary integration (2.92). A simple explanation exists for formula (2.94) in terms of the statistics of classical particles much like the explanation for the Poissonian photon distribution of coherent states. Formula (2.94) appears like a probability distribution of independently produced particle pairs. Photons are generated independently from one another with a probability proportional to $1/2 \tanh \zeta$, that is, proportional to the half of the squeezing parameter ζ for weak pumping. For stronger pumping the generation process becomes saturated. We observe pairs of $2m$ independently produced photons. All photons appear as distinguishable classical particles, but the detector cannot discriminate between them. It detects any m pairs of $2m$ particles, giving rise to a statistical enhancement described by the binomial coefficient in formula (2.94). Note that although this explanation is consistent with the pair statistics (2.94), it loses its meaning when the wave features of light (for instance, interference effects) become important. Figure 2.5 shows an experimental plot of the photon statistics of a squeezed vacuum measured using a method described in detail in Section 5.2.2. We see clearly near-zero probability for odd photon numbers and a decreasing probability for an increasing number of pairs. (The small nonzero probability for odd numbers is caused by detection inefficiencies; see Section 4.1.4.) Because higher-number pairs are produced by higher-order processes, they are less likely to be observed than lower-number pairs. Vacuum has always the lion's share in the photon-number distribution of a squeezed vacuum.

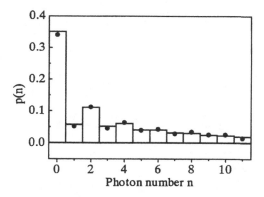

Fig. 2.5. Photon-number distribution of a squeezed vacuum. Photons are produced in pairs. However, detection inefficiencies (see Section 5.3.1) break the pairs so that we observe also odd photon numbers with nonzero probability. The histogram (dots) was reconstructed from measured quadrature distributions via the method described in Section 5.2. The bar chart shows a fit with the theoretical prediction based on Eqs. (2.93) and (2.94) and formula (4.59), taking detection losses into account. [Courtesy of G. Breitenbach, University of Constance.] See also Ref. [240].

Finally, we note that squeezed states are nonclassical states. Because they are pure states and different from coherent states, they cannot be described in terms of statistical mixtures of coherent states. Apart from this trivial formal statement, the quadrature noise reduction below the vacuum level and the photon pairing of squeezed vacuum illustrate beautifully that these states have indeed distinguished quantum properties.

2.4 Further reading

A more complete and detailed introduction to the quantum theory of light is given in most textbooks on quantum optics. For instance, the reader is referred to the books by H.J. Carmichael [55]; C. Cohen-Tannoudji, J. Dupont-Roc, and G. Grynberg [59, 60]; C.W. Gardiner [100]; H. Haken [108]; R. Loudon [178]; W.H. Louisell [181]; P. Meystre and M. Sargent [109]; H.M. Nussenzveig [202]; J. Peřina [217]; W.P. Schleich, D.S. Krähmer, and E. Mayr [245]; M.O. Scully and M.S. Zubairy [251]; D.F. Walls and G.J. Milburn [284]; and W. Vogel and D.-G. Welsch [281] and to the monumental monograph by L. Mandel and E. Wolf [187]. Various aspects of generalized coherent states are considered in the collection [138] of reprinted papers edited by J.R. Klauder and B.-S. Skagerstam. See also the book by A.M. Perelomov [216]. Squeezed states are reviewed by R. Loudon and P.L. Knight [179].

3

Quasiprobability distributions

3.1 Wigner representation

In classical optics the state of an electromagnetic oscillator is perfectly described by the statistics of the classical amplitude α. The amplitude may be completely fixed (then the field is coherent), or α may fluctuate (then the field is partially coherent or incoherent). In classical optics as well as in classical mechanics, we can characterize the statistics of the complex amplitude α or, equivalently, the statistics of the components position q and momentum p introducing a phase-space distribution $W(q, p)$. (As explained in Section 2.1, the real and the imaginary part of the complex amplitude α can be regarded as the position and the momentum of the electromagnetic oscillator.) The distribution $W(q, p)$ quantifies the probability of finding a particular pair of q and p values in their simultaneous measurement. Knowing the phase-space probability distribution, all statistical quantities of the electromagnetic oscillator can be predicted by calculation. In this sense the phase-space distribution describes the state in classical physics. All this is much more subtle in quantum mechanics. First of all, Heisenberg's uncertainty principle prevents us from observing position and momentum simultaneously *and* precisely. So it seems there is no point in thinking about quantum phase space. But wait! In quantum mechanics we cannot directly observe quantum states either. Nevertheless, we are legitimately entitled to use the concept of states as if they were existing entities (whatever they are). We use their properties to predict the statistics of observations. Why not use a quantum phase-space distribution $W(q, p)$ solely to calculate observable quantities in a classicallike fashion? Clearly, the concept of quantum phase space must contain a certain flaw. The probability distribution $W(q, p)$ could become negative, for instance, or ill-behaved. Also, the classicallike the fashion of making statistical predictions may seem to be classicallike at the first glance but not at the second. For these very reasons we should call $W(q, p)$ a *quasiprobability distribution*. Furthermore, there are certainly infinitely many

37

ways of making up quasiprobability distributions (simply because there is no way of defining them properly). Which one shall we choose? Is there a royal road to quantum phase space?

3.1.1 Wigner's formula

Bertrand and Bertrand [29] had the brilliant idea of defining the quasiprobability distribution $W(q, p)$ by postulating its properties. Just one postulate turns out to be sufficient. Let us assume that the distribution $W(q, p)$ *behaves* like a joint probability distribution for q and p without ever mentioning any simultaneous observation of position and momentum. What can we say about classical probability distributions? The *marginal distributions* or, in other words, the reduced distributions $\int_{-\infty}^{+\infty} W(q, p) \, dp$ or $\int_{-\infty}^{+\infty} W(q, p) \, dq$ must yield the position or the momentum distribution, respectively. Additionally, if we perform a phase shift θ all complex amplitudes are shifted in phase, meaning that the components q and p rotate in the two-dimensional phase space (q, p). A classical probability distribution for position and momentum values would rotate accordingly. In view of this fact we postulate that the position probability distribution $\mathrm{pr}(q, \theta)$ after an arbitrary phase shift θ should equal

$$\mathrm{pr}(q, \theta) \equiv \langle q | \hat{U}(\theta) \hat{\rho} \hat{U}^{\dagger}(\theta) | q \rangle$$
$$= \int_{-\infty}^{+\infty} W(q \cos \theta - p \sin \theta, q \sin \theta + p \cos \theta) \, dp. \quad (3.1)$$

This single formula marries the quasiprobability distribution $W(q, p)$ with quantum mechanics. The same formula ties $W(q, p)$ to observable quantities. And, even more remarkably, the formula links quantum states to observations. Considering special cases of formula (3.1) we see that the marginal distributions of $W(q, p)$ produce the correct position and momentum probabilities, respectively. For $\theta = 0$ we obtain

$$\int_{-\infty}^{+\infty} W(q, p) \, dp = \langle q | \hat{\rho} | q \rangle \quad (3.2)$$

and for $\theta = \pi/2$

$$\langle p | \hat{\rho} | p \rangle = \int_{-\infty}^{+\infty} W(-q, p) \, dq = \int_{-\infty}^{+\infty} W(q, p) \, dq. \quad (3.3)$$

[To justify Eq. (3.3) we note that $\hat{U}^{\dagger}(\pi/2) | q \rangle$ is a momentum eigenstate $| p = q \rangle$ with eigenvalue q according to Eq. (2.15).] Integrals such as (3.1) are called *Radon transformations* [226], and they are thoroughly studied in the mathematical theory of tomographic imaging [113], [194]. The inversion of the Radon transformation [226] plays a distinguished role in tomography. In quantum-state tomography the *inverse Radon transformation* turns out to be the mathematical key for quantum-state reconstruction. See Section 5.1.1.

Why is postulate (3.1) sufficient? To understand the reason we introduce the Fourier-transformed distribution $\tilde{W}(u, v)$ called the *characteristic function*

$$\tilde{W}(u, v) \equiv \int_{-\infty}^{+\infty} \int_{-\infty}^{+\infty} W(q, p) \exp(-iuq - ivp)\, dq\, dp \qquad (3.4)$$

and the Fourier-transformed position probability distribution $\widetilde{\mathrm{pr}}(\xi, \theta)$

$$\widetilde{\mathrm{pr}}(\xi, \theta) \equiv \int_{-\infty}^{+\infty} \mathrm{pr}(q, \theta) \exp(-i\xi q)\, dq. \qquad (3.5)$$

On the other hand, the basic postulate (3.1) for $W(q, p)$ requires that

$$\widetilde{\mathrm{pr}}(\xi, \theta) = \int_{-\infty}^{+\infty} \int_{-\infty}^{+\infty} W(q', p') \exp(-i\xi q)\, dq\, dp, \qquad (3.6)$$

with the abbreviations

$$q' = q \cos\theta - p \sin\theta \quad \text{and} \quad p' = q \sin\theta + p \cos\theta. \qquad (3.7)$$

Consequently, q is given by

$$q = q' \cos\theta + p' \sin\theta. \qquad (3.8)$$

We change the integration variables from (q', p') to (q, p) and obtain, according to the very definition of the characteristic function (3.4),

$$\widetilde{\mathrm{pr}}(\xi, \theta) = \tilde{W}(\xi \cos\theta, \xi \sin\theta). \qquad (3.9)$$

The Fourier-transformed position probability distribution is the characteristic function in polar coordinates. In this way the two functions are intimately related. So far we have used only the second line of the fundamental postulate (3.1) or, so to say, the classical nature of the quasiprobability distribution $W(q, p)$. The quantum features come into play when we substitute the first line, that is, the definition of $\mathrm{pr}(q, \theta)$ in the Fourier transformation (3.5). We obtain explicitly

$$\begin{aligned}
\widetilde{\mathrm{pr}}(\xi, \theta) &= \int_{-\infty}^{+\infty} \langle q | \hat{U}(\theta) \hat{\rho} \hat{U}^\dagger(\theta) | q \rangle \exp(-i\xi q)\, dq \\
&= \int_{-\infty}^{+\infty} \langle q | \hat{U}(\theta) \hat{\rho} \hat{U}^\dagger(\theta) \exp(-i\xi \hat{q}) | q \rangle\, dq \\
&= \mathrm{tr}\big\{ \hat{U}(\theta) \hat{\rho} \hat{U}^\dagger(\theta) \exp(-i\xi \hat{q}) \big\} \\
&= \mathrm{tr}\big\{ \hat{\rho} \hat{U}^\dagger(\theta) \exp(-i\xi \hat{q}) \hat{U}(\theta) \big\}.
\end{aligned} \qquad (3.10)$$

We use the rotation formula (2.15) for the quadrature operators to get

$$\hat{U}^\dagger(\theta) \exp(-i\xi \hat{q}) \hat{U}(\theta) = \exp(-i\hat{q}\xi \cos\theta - i\hat{p}\xi \sin\theta). \qquad (3.11)$$

This is the *Weyl operator* [289] $\exp(-iu\hat{q} - iv\hat{p})$ in polar coordinates. The Fourier transform $\widetilde{\mathrm{pr}}(\xi, \theta)$ gives the characteristic function in polar coordinates, as we have learned from Eq. (3.9). Consequently,

$$\tilde{W}(u, v) = \mathrm{tr}\{\hat{\rho} \exp(-iu\hat{q} - iv\hat{p})\}. \tag{3.12}$$

The characteristic function is the "quantum Fourier transform" of the density operator. Because the characteristic function $\tilde{W}(u, v)$ is the Fourier transform of $W(q, p)$ by definition, the quasiprobability distribution $W(q, p)$ should be very closely related to the density operator. Indeed, both are one-to-one representations of the quantum state, as we will show in the next subsection, 3.1.2. Let us calculate the trace in Eq. (3.12) in the position representation. We use the Baker–Hausdorff formula (2.55) to reexpress the Weyl operator

$$\exp(-iu\hat{q} - iv\hat{p}) = \exp\left(-i\frac{uv}{2}\right) \exp(-iu\hat{q}) \exp(-iv\hat{p}). \tag{3.13}$$

The operator $\exp(-iv\hat{p})$ shifts the position eigenstates $|q\rangle$ by v to produce $|q + v\rangle$ because of the relation (2.21) between the position and momentum eigenstates. Consequently,

$$\begin{aligned}
\tilde{W}(u, v) &= \int_{-\infty}^{+\infty} \langle q|\hat{\rho} \exp(-iu\hat{q} - iv\hat{p})|q\rangle \, dq \\
&= \exp\left(-i\frac{uv}{2}\right) \int_{-\infty}^{+\infty} \langle q|\hat{\rho} \exp(-iuq)|q + v\rangle \, dq. \tag{3.14}
\end{aligned}$$

We replace q by $x - v/2$ and obtain the compact formula

$$\tilde{W}(u, v) = \int_{-\infty}^{+\infty} \exp(-iux) \left\langle x - \frac{v}{2}\middle|\hat{\rho}\middle|x + \frac{v}{2}\right\rangle dx. \tag{3.15}$$

To derive an explicit expression for the quasiprobability distribution $W(q, p)$, we simply invert the Fourier transformation in definition (3.4) and get by virtue of formula (3.15) for the characteristic function

$$\begin{aligned}
W(q, p) &= \frac{1}{(2\pi)^2} \int_{-\infty}^{+\infty} \int_{-\infty}^{+\infty} \tilde{W}(u, v) \exp(iuq + ivp) \, du \, dv \\
&= \frac{1}{(2\pi)^2} \int_{-\infty}^{+\infty} \int_{-\infty}^{+\infty} \int_{-\infty}^{+\infty} \left\langle q' - \frac{v}{2}\middle|\hat{\rho}\middle|q' + \frac{v}{2}\right\rangle \\
&\quad \times \exp(-iuq' + iuq + ivp) \, dq' \, du \, dv \\
&= \frac{1}{2\pi} \int_{-\infty}^{+\infty} \int_{-\infty}^{+\infty} \left\langle q' - \frac{v}{2}\middle|\hat{\rho}\middle|q' + \frac{v}{2}\right\rangle \exp(ivp) \\
&\quad \times \delta(q' - q) \, dv \, dq'. \tag{3.16}
\end{aligned}$$

Setting $x = v$ we obtain, finally,

$$W(q, p) = \frac{1}{2\pi} \int_{-\infty}^{+\infty} \exp(\mathrm{i}px) \left\langle q - \frac{x}{2} \middle| \hat{\rho} \middle| q + \frac{x}{2} \right\rangle dx. \qquad (3.17)$$

This is Wigner's legendary formula [290] for a classicallike phase-space distribution in quantum mechanics called the *Wigner function*. It appeared for the first time in his 1932 paper [290] "On the Quantum Correction for Thermodynamic Equilibrium." It was "chosen from all possible expressions, because it seems to be the simplest."*

3.1.2 Basic properties

Wigner's representation of quantum mechanics has found many applications in broad areas of quantum physics. It was used whenever quantum corrections to classical laws were of interest, as in Wigner's 1932 paper [290]. Countless articles have been written on the Wigner function itself. Because there is room neither for mentioning all of them here nor for deriving all known properties of the Wigner function, the reader is referred to Moyal's early yet excellent text [191] and to the more recent review articles by Tatarskii [263]; Balazs and Jennings [10]; Hillery, O'Connell, Scully, and Wigner [116]; and Lee [149] and to the book [135] by Kim and Noz. Here we will focus on only the basic properties of the Wigner function. First, we note that the Wigner function is real

$$W^*(q, p) = W(q, p) \qquad (3.18)$$

for Hermitian operators $\hat{\rho}$. This property is verified by considering the complex conjugate of Wigner's formula (3.17) and replacing x by $-x$. The Wigner function is normalized

$$\int_{-\infty}^{+\infty} \int_{-\infty}^{+\infty} W(q, p) \, dq \, dp = 1, \qquad (3.19)$$

as is easily seen from Wigner's formula (3.17), because the density operator $\hat{\rho}$ is normalized so that $\mathrm{tr}\{\hat{\rho}\} = 1$. So far the Wigner function shows features of a proper probability distribution.

A remarkable property of the Wigner representation is the *overlap formula*

$$\mathrm{tr}\{\hat{F}_1 \hat{F}_2\} = 2\pi \int_{-\infty}^{+\infty} \int_{-\infty}^{+\infty} W_1(q, p) W_2(q, p) \, dq \, dp \qquad (3.20)$$

for the Wigner functions W_1 and W_2 of two arbitrary operators $\hat{F}_1$ and $\hat{F}_2$. Both operators are not even required to be Hermitian, and we have used Wigner's

---

*A footnote, however, says that "this expression was found by L. Szilard and [E.P. Wigner] some years ago for another purpose . . . ."

formula (3.17) with $\hat{F}_k$ instead of $\hat{\rho}$ to define a Wigner function for the $\hat{F}_k$ operators. The proof of the overlap formula (3.20) is a straightforward calculation using Wigner's expression (3.17) for the right-hand side

$$
\frac{1}{2\pi} \int_{-\infty}^{+\infty} \int_{-\infty}^{+\infty} \int_{-\infty}^{+\infty} \int_{-\infty}^{+\infty} \exp[ip(x_1 + x_2)] \left\langle q - \frac{x_1}{2} \middle| \hat{F}_1 \middle| q + \frac{x_1}{2} \right\rangle
$$

$$
\times \left\langle q - \frac{x_2}{2} \middle| \hat{F}_2 \middle| q + \frac{x_2}{2} \right\rangle dx_1 \, dx_2 \, dq \, dp
$$

$$
= \int_{-\infty}^{+\infty} \int_{-\infty}^{+\infty} \left\langle q - \frac{x}{2} \middle| \hat{F}_1 \middle| q + \frac{x}{2} \right\rangle \left\langle q + \frac{x}{2} \middle| \hat{F}_2 \middle| q - \frac{x}{2} \right\rangle dq \, dx
$$

$$
= \int_{-\infty}^{+\infty} \int_{-\infty}^{+\infty} \langle q'|\hat{F}_1|q''\rangle \langle q''|\hat{F}_2|q'\rangle \, dq' \, dq''
$$

$$
= \int_{-\infty}^{+\infty} \langle q'|\hat{F}_1 \hat{F}_2|q'\rangle \, dq'
$$

$$
= \mathrm{tr}\{\hat{F}_1 \hat{F}_2\}. \tag{3.21}
$$

Why is the overlap formula remarkable? We can use it for calculating expectation values

$$
\mathrm{tr}\{\hat{\rho}\hat{F}\} = 2\pi \int_{-\infty}^{+\infty} \int_{-\infty}^{+\infty} W(q, p) W_F(q, p) \, dq \, dp. \tag{3.22}
$$

(We have simply replaced $\hat{F}_1$ by $\hat{\rho}$ and $\hat{F}_2$ by $\hat{F}$.) This equation would be the rule for predicting expectations in classical statistical physics, too. The Wigner function $W(q, p)$ plays the role of a classical phase-space density, whereas $W_F(q, p)$ appears as the physical quantity that is averaged with respect to $W(q, p)$. This is exactly the classicallike fashion of calculating quantum-mechanical expectation values we were seeking. We can understand formula (3.22) another way by seeing $W_F(q, p)$ as a filter function. Consequently, all that quantum mechanics allows us to predict are filtered projections of the Wigner function. All that we can see are shadows of the states, very much in the sense of Plato's famous parable [219] mentioned in the introduction, yet formulated more precisely, that is, more quantitatively.

Another simple consequence of the overlap formula (3.20) is the expression

$$
|\langle \psi_1 \mid \psi_2 \rangle|^2 = 2\pi \int_{-\infty}^{+\infty} \int_{-\infty}^{+\infty} W_1(q, p) W_2(q, p) \, dq \, dp \tag{3.23}
$$

for the transition probability between the pure states $|\psi_1\rangle$ and $|\psi_2\rangle$. However, this quantity vanishes if the states $|\psi_1\rangle$ and $|\psi_2\rangle$ are orthogonal

$$
\langle \psi_1 \mid \psi_2 \rangle = 0. \tag{3.24}
$$

The overlap (3.23) of two positive functions W_1 and W_2 cannot be zero. Consequently, Wigner functions cannot be positive in general. (In fact, states having Gaussian wave functions are the only pure states with nonnegative Wigner functions. See [119], [185], and [263].) In this way the overlap formula (3.20) reveals strikingly both the similarities and the differences between a classical probability distribution and the Wigner function. Quantum interference implies that the classicallike Wigner function cannot be regarded as a probability distribution but as a quasiprobability distribution only. This property is one way in which the unavoidable flaw in the concept of quantum phase space may appear. Negative regions in the Wigner function of a given state can be seen as signatures of nonclassical behavior [185].

Using the overlap formula (3.20) we are also able to quantify the purity of a quantum state. In fact, identifying both $\hat{F}_1$ and $\hat{F}_2$ with $\hat{\rho}$ we obtain

$$\mathrm{tr}\{\hat{\rho}^2\} = 2\pi \int_{-\infty}^{+\infty} \int_{-\infty}^{+\infty} W(q, p)^2 \, dq \, dp. \qquad (3.25)$$

The *purity* $\mathrm{tr}\{\hat{\rho}^2\}$ ranges between zero and unity and equals exactly unity if and only if the state is pure ($\hat{\rho} = |\psi\rangle\langle\psi|$). According to relation (1.21) the von-Neumann entropy S is bounded by

$$S \equiv -\mathrm{tr}\{\hat{\rho} \ln \hat{\rho}\} \geq 1 - 2\pi \int_{-\infty}^{+\infty} \int_{-\infty}^{+\infty} W(q, p)^2 \, dq \, dp. \qquad (3.26)$$

We see that the overlap of the Wigner function with itself provides a convenient way of expressing statistical purity in quantum mechanics.

Finally, we can use the overlap formula (3.20) to represent the density-matrix elements in a given basis in terms of the Wigner function

$$\langle a'|\hat{\rho}|a\rangle = \mathrm{tr}\{\hat{\rho}|a\rangle\langle a'|\} = 2\pi \int_{-\infty}^{+\infty} \int_{-\infty}^{+\infty} W(q, p) W_{a'a}(q, p) \, dq \, dp. \qquad (3.27)$$

Here $W_{a'a}(q, p)$ denotes the Wigner representation of the projector $|a\rangle\langle a'|$ obtained by replacing $\hat{\rho}$ by $|a\rangle\langle a'|$ in Wigner's formula (3.17). This property shows that the Wigner function is indeed a one-to-one representation of the quantum state.

We may turn the tables and ask, is any normalized real function $W(q, p)$ always a Wigner function, that is, does it correspond to a state? Obviously it does not, because the integral of the squared function must be less than or equal to $(2\pi)^{-1}$ according to the purity relation (3.25). Another quantum constraint is imposed on any realistic Wigner function. The values of it may range between only $\pm\pi^{-1}$, that is

$$|W(q, p)| \leq \frac{1}{\pi}. \qquad (3.28)$$

To prove this inequality we consider a pure state $\hat{\rho} = |\psi\rangle\langle\psi|$ first. We use the Schwarz inequality to estimate the Wigner function given in terms of Wigner's formula (3.17) and obtain

$$|W(q,p)|^2 \le \frac{1}{(2\pi)^2} \int_{-\infty}^{+\infty} \left|\left\langle q - \frac{x}{2}\middle|\psi\right\rangle\right|^2 dx \int_{-\infty}^{+\infty} \left|\left\langle q + \frac{x}{2}\middle|\psi\right\rangle\right|^2 dx = \frac{1}{\pi^2}$$

(3.29)

because the state vector $|\psi\rangle$ is normalized. In case of a statistical mixture the density matrix can be represented as a sum of pure states $|\psi_a\rangle\langle\psi_a|$ weighted by their probabilities ρ_a according to the very definition (1.12) of the density operator. Consequently, the Wigner function for a mixed state is a weighted sum of pure Wigner functions as well. By estimating the individual pure Wigner functions and summing with respect to the normalized probabilities p_a, we see easily that the bound (3.28) is valid for mixed states, too. [Note that the Wigner function $W_n(q,p)$ for Fock states $|n\rangle$ actually equals $(-1)^n/\pi$ at the origin $q = p = 0$; see Eq. (3.83).] The constraint (3.28) shows that Wigner functions cannot be highly peaked, meaning that the quantum "phase-space density" cannot be arbitrarily high and Wigner functions cannot approach delta functions $\delta(q - q_0)\delta(p - p_0)$, for instance. Of course, according to Heisenberg's uncertainty principle, position and momentum must fluctuate statistically, and this intrinsic uncertainty is mirrored in the uniform bound (3.28). Note that other constraints on Wigner functions were given by Tatarskii [263] and Lieb [175]. However, no golden rule decides whether a given function is a Wigner function, apart from the Solomonic statement that any density matrix derived from a proper Wigner function should be a density matrix, or have nonnegative eigenvalues. Equivalently, all main-diagonal elements $\langle a|\hat{\rho}|a\rangle$ derived according to formula (3.27) must be nonnegative. Deviations from this law indicate imperfections in experimentally reconstructed Wigner functions, for instance.

Apart from formula (3.22) another equivalent way exists of making quantum-mechanical predictions, that is, of calculating expectation values via the Wigner function. We consider

$$\text{tr}\{\hat{\rho}(\lambda\hat{q} + \mu\hat{p})^k\} = i^k \frac{\partial^k}{\partial\xi^k}\tilde{W}(\xi\lambda, \xi\mu)\Big|_{\xi=0}$$

$$= \int_{-\infty}^{+\infty}\int_{-\infty}^{+\infty} W(q,p)(\lambda q + \mu p)^k\, dq\, dp.$$ (3.30)

In the first line we have used key formula (3.12) for the characteristic function $\tilde{W}(u,v)$, whereas in the second line we have used the Fourier relationship (3.4) between $\tilde{W}(u, v)$ and the Wigner function. Comparing the powers of λ and μ we see that

$$\text{tr}\{\hat{\rho}\mathcal{S}\hat{q}^m\hat{p}^n\} = \int_{-\infty}^{+\infty}\int_{-\infty}^{+\infty} W(q, p)q^m p^n \, dq \, dp. \qquad (3.31)$$

The symbol $\mathcal{S}$ means that we should symmetrize all possible products of the m $\hat{q}$-operators and the n $\hat{p}$-operators, that is, we should take the average over all products with the right amount of $\hat{q}$'s and $\hat{p}$'s. So, for example, $1/3$ $(\hat{q}^2\hat{p} + \hat{q}\hat{p}\hat{q} + \hat{p}\hat{q}^2)$ corresponds to q^2p. This *Weyl correspondence* [289] is also a convenient way of making quantum-mechanical predictions in a classicallike fashion. [Note that the Weyl correspondence is completely equivalent to formula (3.22).] Given a symmetrized operator $\hat{F}$, we can calculate quantum-mechanical averages as if $\hat{F}$ were a classical quantity. However, this pleasing property is Janus-faced. The square of $\hat{F}$, which describes the fluctuations of $\hat{F}$, is not necessarily symmetrized [and the Wigner function of $\hat{F}^2$ is not always $W_F(q, p)^2$]. We should express $\hat{F}^2$ in terms of symmetrized operators to get meaningful results. This route is another way in which the mutual exclusion of certain observables sneaks in via the commutator relations of position and momentum. So we must not forget that the algebraic structures of quantum mechanics and classical physics are different, despite many similarities. This difference causes a problem in the very concept of a quantum phase space even more serious than negative "probabilities."

3.1.3 Examples

How do typical Wigner functions look? Are they similar to classical phase-space densities? Probably the simplest example is the Wigner function for the vacuum state. We insert the quadrature wave function (2.33) in Wigner's formula (3.17) and see that the Wigner function for a vacuum is Gaussian

$$W_0(q, p) = \frac{1}{\pi}\exp(-q^2 - p^2). \qquad (3.32)$$

Classically, this function would correspond to the phase-space density of an ensemble of electromagnetic oscillators fluctuating statistically around the origin in phase space with isotropic variances of $1/2$ in our units. Quantum-mechanically, these statistical fluctuations occur even if the spatial–temporal mode is in a pure vacuum. Figure 3.1 shows the experimentally reconstructed Wigner function for a vacuum, illustrating beautifully the isotropic character of the vacuum fluctuations (except for small experimental errors). How does a

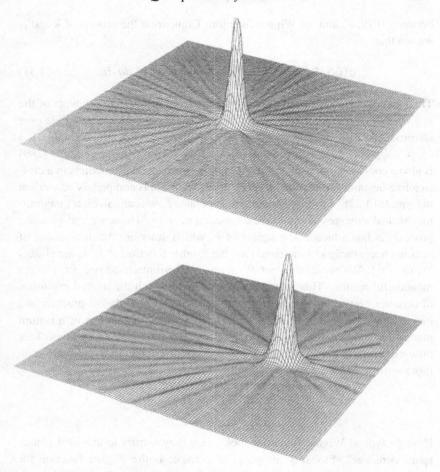

Fig. 3.1. Wigner function for a vacuum (top) and for a coherent state (bottom). We see clearly that coherent states are just "displaced vacuums." Optical homodyne tomography (see Chapter 5) was used to reconstruct the Wigner functions from experimental data. [Courtesy of G. Breitenbach, University of Constance.]

squeezed vacuum look? Let us study the general effect of squeezing in phase space first. We obtain from Wigner's formula (3.17)

$$
\begin{aligned}
W_s(q, p) &= \frac{1}{2\pi} \int_{-\infty}^{+\infty} \exp(\mathrm{i}px) \left\langle q - \frac{x}{2} \middle| \hat{S}\hat{\rho}\hat{S}^\dagger \middle| q + \frac{x}{2} \right\rangle dx \\
&= \frac{1}{2\pi} \int_{-\infty}^{+\infty} \exp(\mathrm{i}px) \mathrm{e}^\zeta \left\langle \mathrm{e}^\zeta \left(q - \frac{x}{2} \right) \middle| \hat{\rho} \middle| \mathrm{e}^\zeta \left(q + \frac{x}{2} \right) \right\rangle dx \quad (3.33)
\end{aligned}
$$

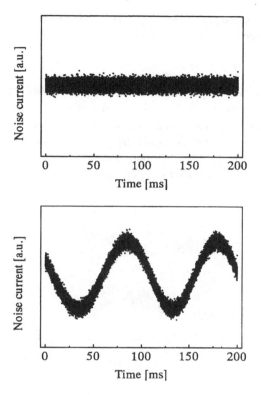

Fig. 3.2. Quadrature fluctuations of a vacuum (top) and a coherent state (bottom). Homodyne detection (see Section 4.2) was employed for performing the quadrature measurements on stationary fields. Each point represents a single measurement result. The reference phase (the phase of the local oscillator) was gradually shifted during the observation time. We see clearly the phase independence of the vacuum fluctuations and the oscillating effect, Eq. (2.47), of phase shifts on the quadratures of a coherent state. The plot shows a part of the experimental data used to reconstruct the Wigner functions of the previous figure via optical homodyne tomography (see Chapter 5). [Courtesy of G. Breitenbach, University of Constance.]

because the squeezing operator $\hat{S}$ defined in Eq. (2.82) rescales the position wave function according to Eq. (2.79). We substitute $e^{\zeta} x$ with x' and get the result

$$W_s(q, p) = W(e^{\zeta} q, e^{-\zeta} p). \qquad (3.34)$$

The Wigner function for a squeezed state is squeezed in one quadrature direction and stretched accordingly in the orthogonal line in order to preserve the area in phase space. In this way the quadrature fluctuations displayed in the Wigner function are redistributed from one quadrature to the canonically conjugate quantity. This redistribution is exactly what we would expect from squeezing in

phase space. Using the general result (3.34), we obtain directly from Eq. (3.32)
the Wigner function of a squeezed vacuum

$$W_s(q, p) = \frac{1}{\pi} \exp(-e^{2\zeta} q^2 - e^{-2\zeta} p^2). \qquad (3.35)$$

As for a vacuum, the Wigner function is a Gaussian distribution with, however, unbalanced variances (2.78) indicating the effect of quadrature squeezing.
Figure 3.3 shows the experimentally reconstructed Wigner function of a significantly squeezed vacuum generated by parametric amplification. It is an
easy exercise to calculate the distribution $\mathrm{pr}(q, \theta)$ for phase-shifted quadratures from the Wigner function (3.35) via the Radon transformation (3.1). We

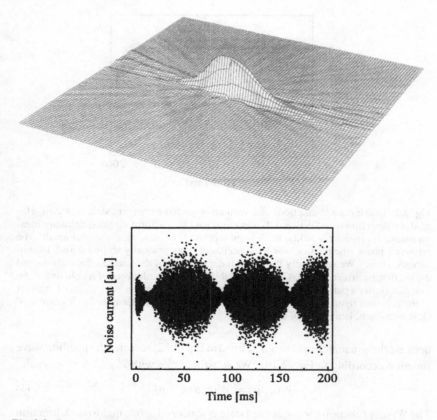

Fig. 3.3. Squeezed vacuum. Wigner function (top) and quadrature fluctuations (bottom).
We see clearly a remarkable squeezing in phase space and the corresponding breathing
of the quadrature noise, Eq. (3.37). The noise trace shows a part of the experimental data
used to reconstruct the depicted Wigner function via optical homodyne tomography (see
Chapter 5). [Courtesy of G. Breitenbach, University of Constance.] See also Ref. [41].

find the result

$$\text{pr}(q, \theta) = \left(2\pi \Delta_\theta^2 q\right)^{-1/2} \exp\left(-\frac{q^2}{2\Delta_\theta^2 q}\right) \tag{3.36}$$

with the phase-dependent variance

$$\Delta_\theta^2 q = \frac{1}{2}(e^{2\varsigma} \sin^2 \theta + e^{-2\varsigma} \cos^2 \theta). \tag{3.37}$$

The quadrature fluctuations of a squeezed vacuum are Gaussian and, of course, phase dependent. Their variances $\Delta_\theta^2 q$ vary from $\frac{1}{2}e^{-2\varsigma}$ to $\frac{1}{2}e^{+2\varsigma}$ with a period of π.

What is the Wigner function of a coherent state? Coherent states are displaced vacuums, so we would expect that their Wigner functions are displaced vacuum Wigner functions, too, with a displacement given by the complex coherent amplitude $2^{1/2}\alpha = q_0 + ip_0$. That this expectation is correct is easily seen, considering the general effect of the displacement operator $\hat{D}$ in phase space

$$
\begin{aligned}
W_D(q, p) &= \frac{1}{2\pi} \int_{-\infty}^{+\infty} \exp(ipx)\left\langle q - \frac{x}{2}\middle|\hat{D}\hat{\rho}\hat{D}^\dagger\middle|q + \frac{x}{2}\right\rangle dx \\
&= \frac{1}{2\pi} \int_{-\infty}^{+\infty} \exp[i(p - p_0)x] \\
&\quad \times \left\langle q - \frac{x}{2} - q_0\middle|\hat{\rho}\middle|q + \frac{x}{2} - q_0\right\rangle dx
\end{aligned} \tag{3.38}
$$

according to the general rule (2.56) and (2.59) for displacing position wave functions. We find that Wigner functions of displaced states are indeed just displaced Wigner functions

$$W_D(q, p) = W(q - q_0, p - p_0) \tag{3.39}$$

and, consequently, the Wigner function of a coherent state is given by the displaced Gaussian distribution

$$W(q, p) = \frac{1}{\pi} \exp[-(q - q_0)^2 - (p - p_0)^2]. \tag{3.40}$$

Again, the Wigner function displays the typical features of the considered quantum state: A coherent state as produced by a high-quality laser has a stable coherent amplitude $q_0 + ip_0$ that is contaminated by the unavoidable vacuum fluctuations only.

According to the fundamental superposition principle of quantum mechanics, we are entitled to think of quantum superpositions of coherent states. These are states that contain simultaneously two coherent components, one pointing in one direction in phase space and the other pointing in another. The position wave function ψ of such a state would be the superposition of two coherent-state

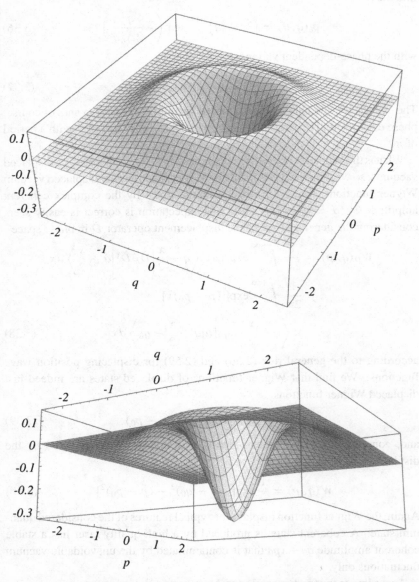

Fig. 3.4. Wigner function of a single photon (a one-photon Fock state) seen from above and from below. According to Eq. (3.83), the Wigner function is given by the expression $W(q, p) = \exp(-q^2 - p^2)(2q^2 + 2p^2 - 1)/\pi$. Negative "probabilities" are clearly visible near the origin of the phase space.

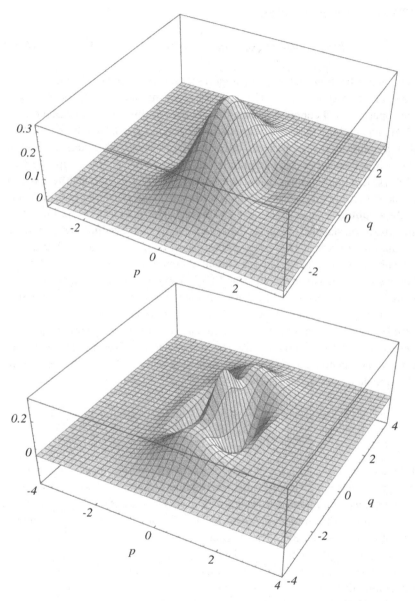

Fig. 3.5. Wigner function of Schrödinger-cat states defined in Eq. (3.41). Top: $q_0 = 1$. The "cat" shows only a mere squeezing instead of a clear separation of two coherent amplitudes. Bottom: $q_0 = 2$. First indications of two distinct humps are visible in the quantum-interference structure.

wave functions; for instance,

$$\psi(q) \propto \exp\left[-\frac{1}{2}(q - q_0)^2\right] + \exp\left[-\frac{1}{2}(q + q_0)^2\right]. \qquad (3.41)$$

(The normalization factor is not important here and has been omitted.) The wave function shows two peaks, one at q_0 and the other at $-q_0$ according to the superimposed coherent amplitudes. Note that this quantum superposition (3.41) has nothing to do with optical interference. When two fields interfere, their amplitude may be enhanced or canceled, producing, for example, coherent states of enhanced or zero amplitude (vacuum). The quantum superposition (3.41) still contains both coherent amplitudes $\pm q_0$. It is also much different from an incoherent superposition of $\pm q_0$, where the field has either the amplitude $+q_0$ or the amplitude $-q_0$ with certain probabilities. The quadrature amplitude of the superposition state (3.41) is $+q_0$ *as well as* $-q_0$, with a resolution given by the vacuum fluctuations. This strange behavior of being at $+q_0$ as well as at $-q_0$ resembles Schrödinger's famous Gedanken experiment about a quantum cat being simultaneously alive and dead [246]. Therefore, states such as (3.41) are named *Schrödinger-cat states*. They have not yet been observed in the optical domain, because they are extremely vulnerable to losses. Quantum decoherence [308] caused by linear losses is the main reason that extremely strange quantum phenomena allowed in theory are very difficult to observe in practice. Which observable phenomena of the Schrödinger-cat state (3.41) would we expect? We calculate the Wigner function using Wigner's formula (3.17) and obtain

$$W(q, p) \propto \exp[-(q - q_0)^2 - p^2] + \exp[-(q + q_0)^2 - p^2]$$
$$+ 2\exp(-q^2 - p^2)\cos(2pq_0). \qquad (3.42)$$

Like the wave function, the Wigner function exhibits two peaks at the coherent amplitudes $\pm q_0$. However, the interference structure halfway between the peaks displays the quantum superposition of both amplitudes, showing rapid oscillations with a frequency given by the distance $2q_0$ of the superimposed amplitudes. See Fig. 3.6. The Wigner function becomes negative, indicating the nonclassical behavior of the Schrödinger-cat state [47], [243]. To predict observable effects of the quantum-superposition state (3.41) we calculate the quadrature distributions $\mathrm{pr}(q, \theta)$ via Radon transformation (3.1) of the Wigner function (3.42) and get

$$\mathrm{pr}(q, \theta) \propto \exp[-(q - q_0\cos\theta)^2] + \exp[-(q + q_0\cos\theta)^2]$$
$$+ 2\exp\left(-q^2 - q_0^2\cos^2\theta\right)\cos(2qq_0\sin\theta). \qquad (3.43)$$

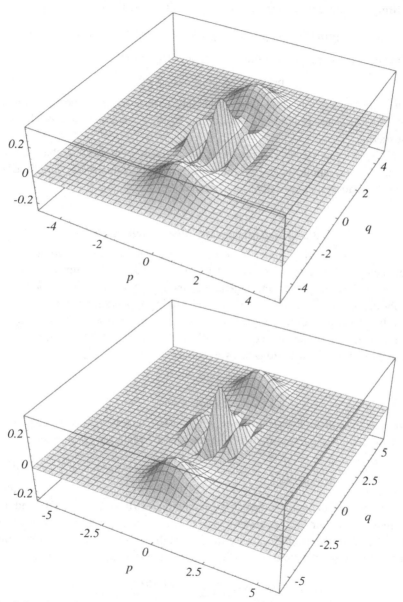

Fig. 3.6. Wigner function of Schrödinger-cat states defined in Eq. (3.41). Top: $q_0 = 3$. Two separated coherent amplitudes are clearly visible. Bottom: $q_0 = 4$. The larger the separation of the amplitudes, the more rapid is the oscillation in the quantum-interference structure.

Shifting the phase θ turns the position quadrature distribution

$$\text{pr}(q) \approx \frac{1}{2} \exp[-(q - q_0)^2] + \frac{1}{2} \exp[-(q + q_0)^2] \qquad (3.44)$$

(showing peaks at $\pm q_0$) into the momentum distribution

$$\text{pr}(p) \propto \exp(-p^2) \cos^2(pq_0) \qquad (3.45)$$

at $\theta = \pi/2$ and $q = p$, displaying typical interference fringes. The interference pattern is mirrored in the highly oscillating Wigner function of a Schrödinger-cat state.

We have seen that Wigner functions are useful to visualize the phase-space properties of quantum states. Wigner functions display quadrature amplitudes, their fluctuations, and possible interferences.

3.2 Other quasiprobability distributions

In many respects the Wigner representation appears as the best compromise between a classical phase-space density and the correct quantum-mechanical behavior. The Wigner function generates the right marginal distributions and it obeys the overlap relation (3.20) for calculating expectation values in a classicallike fashion. Yet the Wigner function may be negative. Is there a way to define a strictly nonnegative quasiprobability distribution? Which other useful quasiprobability distributions can we define?

3.2.1 Q function

We may smooth the Wigner function $W(q, p)$ by convolving it with a Gaussian distribution having the same width as vacuum to obtain the *Q function*

$$Q(q, p) \equiv \frac{1}{\pi} \int_{-\infty}^{+\infty} \int_{-\infty}^{+\infty} W(q', p') \exp[-(q - q')^2 - (p - p')^2] \, dq' \, dp'. \qquad (3.46)$$

What does this expression mean? We recall the overlap relation (3.20) and the formula (3.40) for Wigner functions of coherent states. We see immediately that the Q function gives simply the probability distribution for finding the coherent states $|\alpha\rangle$ with $\alpha = 2^{-1/2}(q + ip)$ in the state $\hat{\rho}$, because

$$Q(q, p) = \frac{1}{2\pi} \text{tr}\{\hat{\rho}|\alpha\rangle\langle\alpha|\}$$

$$= \frac{1}{2\pi} \langle\alpha|\hat{\rho}|\alpha\rangle. \qquad (3.47)$$

(Note also that $\pi^{-1}\langle\alpha|\hat{\rho}|\alpha\rangle$ is frequently called the Q function.) Clearly, the Q function is nonnegative and normalized to unity, as is easily seen from the completeness relation (2.67) of the coherent states. Consequently, the Q function can be regarded as describing probability densities. In fact, we will consider in Chapter 6 a scheme to measure the Q function directly as a probability distribution. We also see that the negative regions of the Wigner function cannot be extended over areas significantly wider than $1/2$; otherwise, the Q function could be negative. (This property puts yet another constraint on Wigner functions.) Roughly speaking, the Gaussian smoothing (3.46) means taking the average of $W(q', p')$ values in a circle area around (q, p) with a radius given by the vacuum fluctuations. Because any negativities disappear after this procedure, they must be concentrated in small regions of the Wigner function. The resolution of these negative "probabilities" requires an accuracy in the order of the vacuum fluctuations.

The smoothing of the Wigner function is also clearly seen in the Fourier-transformed Q function $\tilde{Q}(u, v)$. In fact, we obtain from the definition (3.46)

$$\tilde{Q}(u, v) \equiv \int_{-\infty}^{+\infty} \int_{-\infty}^{+\infty} Q(q, p) \exp(-iuq - ivp) \, dq \, dp \qquad (3.48)$$

$$= \tilde{W}(u, v) \exp\left[-\frac{1}{4}(u^2 + v^2)\right]. \qquad (3.49)$$

Because details of the Wigner function correspond to high-frequency components (u, v), these details are suppressed in the Q representation.

Eq. (3.49) reveals also another important property of the Q function. We use the formula (3.12) for the characteristic function $\tilde{W}(u, v)$ and obtain

$$\tilde{Q} = \text{tr}\left\{\hat{\rho} \exp\left[-\frac{1}{4}(u^2 + v^2)\right] \exp(-iu\hat{q} - iv\hat{p})\right\} \qquad (3.50)$$

or, introducing the complex notation $\beta = 2^{-1/2}(u + iv)$,

$$\tilde{Q} = \text{tr}\left\{\hat{\rho} \exp\left(-\frac{1}{2}|\beta|^2\right) \exp(-i\hat{a}\beta^* - i\hat{a}^\dagger\beta)\right\} \qquad (3.51)$$

because $\hat{a} = 2^{-1/2}(\hat{q} + i\hat{p})$. Finally, we employ the Baker–Hausdorff formula (2.55) and get

$$\tilde{Q} = \text{tr}\left\{\hat{\rho} \exp(-i\hat{a}\beta^*) \exp\left(-i\hat{a}^\dagger\beta\right)\right\}. \qquad (3.52)$$

Consequently,

$$\text{tr}\left\{\hat{\rho}\hat{a}^\nu\hat{a}^{\dagger\mu}\right\} = i^{\nu+\mu} \frac{\partial^\nu}{\partial\beta^{*\nu}} \frac{\partial^\mu}{\partial\beta^\mu} \tilde{Q}\bigg|_{\beta=\beta^*=0}$$

$$= \int_{-\infty}^{+\infty} \int_{-\infty}^{+\infty} Q(q, p)\alpha^\nu\alpha^{*\mu} \, dq \, dp \qquad (3.53)$$

using the definition (3.48) of the Fourier-transformed Q function in the complex notation $\alpha = 2^{-1/2}(q + ip)$. Expectation values of the form $\text{tr}\{\hat{\rho}\hat{a}^\nu\hat{a}^{\dagger\mu}\}$ are called *antinormally ordered*. We have seen that we can express these quantities in terms of the Q function as if $\hat{a}$ and $\hat{a}^\dagger$ were classical amplitudes and not operators. We note, however, that this property relies critically on the ordering $\hat{a}^\nu \hat{a}^{\dagger\mu}$, and it is clearly lost when powers $(\hat{a}^\nu \hat{a}^{\dagger\mu})^\lambda$ are concerned. (Remember the discussion at the end of Section 3.1.2.)

3.2.2 *P function*

In the theory of photodetection (see for instance Ref. [187], Chap. 12), *normally ordered* expectation values $\mathrm{tr}\{\hat{\rho}\hat{a}^{\dagger\mu}\hat{a}^\nu\}$ play a distinguished role. How can we find the phase-space correspondence for normal ordering? We simply reverse the order of the exponentials $\exp(-i\hat{a}\beta^*)$ and $\exp(-i\hat{a}^\dagger\beta)$ in the expression (3.52) to define a new function

$$\tilde{P}(u, v) \equiv \mathrm{tr}\left\{\hat{\rho}\exp\left(-i\hat{a}^\dagger\beta\right)\exp(-i\hat{a}\beta^*)\right\} \tag{3.54}$$

with the convention $\beta = 2^{-1/2}(u + iv)$. Using the same arguments as in the previous subsection we see that the *P function*

$$P(q, p) \equiv \frac{1}{(2\pi)^2}\int_{-\infty}^{+\infty}\int_{-\infty}^{+\infty}\tilde{P}(u, v)\exp(iuq + ivp)\, du\, dv \tag{3.55}$$

corresponds to the normal ordering

$$\mathrm{tr}\{\hat{\rho}\hat{a}^{\dagger\mu}\hat{a}^\nu\} = \int_{-\infty}^{+\infty}\int_{-\infty}^{+\infty}P(q, p)\alpha^{*\mu}\alpha^\nu\, dq\, dp \tag{3.56}$$

with $\alpha = 2^{-1/2}(q + ip)$. Because normally ordered quantities are quite fundamental in quantum optics, the property (3.56) is one reason that the P function (also called the *Glauber–Sudarshan function* [106], [261]) is a rather popular phase-space distribution. Yet another reason is that the P function diagonalizes the density operator in terms of coherent states. To see this property, we argue along similar lines as in the previous subsection where we started from (3.49) and arrived at (3.52). Here we do the necessary algebra in reversed order – we start from the definition (3.54) and arrive via the Baker–Hausdorff formula (2.55) at

$$\tilde{W}(u, v) = \tilde{P}(u, v)\exp\left[-\frac{1}{4}(u^2 + v^2)\right]. \tag{3.57}$$

Consequently,

$$W(q, p) = \int_{-\infty}^{+\infty}\int_{-\infty}^{+\infty}P(q_0, p_0)\frac{1}{\pi}\exp[-(q - q_0)^2 - (p - p_0)^2]\, dq_0\, dp_0. \tag{3.58}$$

We recall the formula (3.40) for Wigner functions of coherent states $|\alpha\rangle$ with $\alpha = 2^{-1/2}(q_0 + ip_0)$, and we use the general correspondence between the

Wigner function and the density matrix to obtain the famous result [261] (called the *optical equivalence theorem* [137])

$$\hat{\rho} = \int_{-\infty}^{+\infty} \int_{-\infty}^{+\infty} P(q_0, p_0) |\alpha\rangle \langle \alpha| \, dq_0 \, dp_0. \tag{3.59}$$

At first glance this formula appears as a representation of the quantum state in terms of a distribution of coherent states, that is, as a statistical mixture of classical amplitudes. This is impossible! There is no way to represent a pure state $|\psi\rangle$ as a mixture of coherent states, unless $|\psi\rangle$ itself is a coherent state. Yet the algebra to arrive at the result (3.59) is correct. What is wrong? The answer is that the P function might be very ill-behaved. For instance, we see from Eq. (3.57) that the Wigner function is a smoothed P function. Because the Wigner function can display negative "probabilities" the P function might behave even worse, that is, it might be negative or it might not even exist as a tempered distribution. (States having such P functions are called *nonclassical states*. See the discussion at the beginning of Section 2.2.3.) Because the P function is such a delicate mathematical construction, there is no practical way to reconstruct it from experimental data.

3.2.3 s-parameterized quasiprobability distributions

We may also convolve the Wigner function with Gaussian distributions having a width different from what would correspond to the vacuum noise. In this way we obtain a whole family of distributions called the *s-parameterized quasiprobability distributions* $W(q, p; s)$ [51], [52], [156]. First, we define the characteristic functions

$$\tilde{W}(u, v; s) \equiv \tilde{W}(u, v) \exp\left[\frac{s}{4}(u^2 + v^2)\right]. \tag{3.60}$$

For historical reasons [51], [52] the real parameter s happens to be negative when the Wigner function is smoothed. We see from definition (3.60) that in this case the distribution is indeed suppressing high frequencies (u, v) describing details in the Wigner function. The s-parameterized quasiprobability distributions are obtained from the characteristic function via inverse Fourier transformation

$$W(q, p; s) \equiv \frac{1}{(2\pi)^2} \int_{-\infty}^{+\infty} \int_{-\infty}^{+\infty} \tilde{W}(u, v; s) \exp(iuq + ivp) \, du \, dv. \tag{3.61}$$

Obviously, all previously studied quasiprobability distributions are included in this family of functions because they correspond to the parameters

$$s = \begin{cases} +1: & P \text{ function}, \\ 0: & \text{Wigner function}, \\ -1: & Q \text{ function}, \end{cases} \tag{3.62}$$

respectively. In this way the defined distributions interpolate between the P, the Wigner, and the Q function. The range of s, however, is the whole real axis. Note that it is also possible to define quasiprobability distributions corresponding to complex s parameters [300].

Let us study some general properties of s-parameterized quasiprobability distributions. Of course, they are normalized to unity because

$$\int_{-\infty}^{+\infty}\int_{-\infty}^{+\infty} W(q, p; s)\, dq\, dp = \tilde{W}(0, 0; s) = 1. \tag{3.63}$$

We obtain from the obvious relation

$$\tilde{W}(u, v; s) = \tilde{W}(u, v; s')\exp\left[\frac{1}{4}(s - s')(u^2 + v^2)\right] \tag{3.64}$$

the formula

$$W(q, p; s) = \frac{1}{\pi(s' - s)}\int_{-\infty}^{+\infty}\int_{-\infty}^{+\infty} W(q', p'; s')$$
$$\times \exp\left[-\frac{(q - q')^2 + (p - p')^2}{(s' - s)}\right] dq'\, dp', \tag{3.65}$$

provided that $s < s'$ so that the integral converges. This relation shows that there is a smoothing hierarchy among the s-parameterized quasiprobability distributions. The smaller the parameter s is, the more the distribution is smoothed. Moreover, the marginal distributions $\mathrm{pr}(q, \theta; s)$ of smoothed Wigner functions ($s < 0$) are smoothed accordingly, that is,

$$\mathrm{pr}(q, \theta; s) \equiv \int_{-\infty}^{+\infty} W(q\cos\theta - p\sin\theta, q\sin\theta + p\cos\theta; s)\, dp \tag{3.66}$$

$$= (\pi|s|)^{-1/2}\int_{-\infty}^{+\infty} \mathrm{pr}(q', \theta)\exp[-|s|^{-1}(q - q')^2]\, dq', \tag{3.67}$$

because the Wigner function has the quantum-mechanically correct marginals $\mathrm{pr}(q, \theta)$. Additionally, the overlap relation (3.20) must be modified for s-parameterized quasiprobability distributions because

$$\mathrm{tr}\{\hat{F}_1\hat{F}_2\} = 2\pi\int_{-\infty}^{+\infty}\int_{-\infty}^{+\infty} W_1(q, p)W_2(q, p)\, dq\, dp$$

$$= \frac{1}{2\pi}\int_{-\infty}^{+\infty}\int_{-\infty}^{+\infty} \tilde{W}_1(u, v)\tilde{W}_2(-u, -v)\, du\, dv$$

$$= \frac{1}{2\pi}\int_{-\infty}^{+\infty}\int_{-\infty}^{+\infty} \tilde{W}_1(u, v; s)\tilde{W}_2(-u, -v; -s)\, du\, dv. \tag{3.68}$$

Consequently, we obtain

$$\mathrm{tr}\{\hat{F}_1\hat{F}_2\} = 2\pi\int_{-\infty}^{+\infty}\int_{-\infty}^{+\infty} W_1(q, p; s)W_2(q, p; -s)\, dq\, dp \tag{3.69}$$

and in particular

$$\text{tr}\{\hat{\rho}\hat{F}\} = 2\pi \int_{-\infty}^{+\infty} \int_{-\infty}^{+\infty} W(q, p; s) W_F(q, p; -s) \, dq \, dp. \tag{3.70}$$

This relation shows that a smoothing of the quasiprobability distribution must be compensated for by an enhancement of the resolution of the filter function $W_F(q, p)$ to calculate expectation values, and vice versa. This procedure may cause significant problems because it requires extremely high accuracy for $W(q, p; s)$ and it may involve singular filter functions $W_F(q, p; s)$. We see that the price to be paid for having a nonnegative quasiprobability distribution is the introduction of additional noise in practical applications. This noise appears in the smoothing of the marginal distributions, and it must be compensated for by enhancing filter functions to correctly predict observable quantities. Finally, we also note that a certain *s-ordering* of operators [52] can be defined to calculate expectation values. However, apart from the normal ordering for the P function, the symmetric ordering for the Wigner representation, and the antinormal ordering corresponding to the Q function, these ordering procedures are involved. The reader is referred to the comprehensive articles by Cahill and Glauber [51], [52] for the details.

3.3 Examples

How do Q functions look? How smoothed are they compared with Wigner functions? How singular can a P function be? Let us study some examples. The simplest candidate to consider theoretically is a Fock state $|n\rangle$. We see immediately from formula (3.47) and the Poissonian photon statistics (2.64) of a coherent state that the Q function of a Fock state is given by

$$\begin{aligned}
Q(q, p) &= \frac{1}{2\pi} |\langle \alpha \mid n \rangle|^2 \\
&= \frac{1}{2\pi n!} \exp(-|\alpha|^2) |\alpha|^{2n} \tag{3.71} \\
&= \frac{1}{2\pi n!} \exp\left[-\frac{1}{2}(q^2 + p^2)\right] \left(\frac{q^2 + p^2}{2}\right)^n. \tag{3.72}
\end{aligned}$$

According to the Gaussian approximation for the Poissonian distribution [279], [241] we can approximate (3.72) for large quantum numbers and get

$$Q_n \sim \frac{1}{2\pi^{3/2} r} \exp[-(r - r_n)^2] \tag{3.73}$$

with

$$r = \sqrt{q^2 + p^2} \tag{3.74}$$

and the Bohr–Sommerfeld radius [79]

$$r_n = \sqrt{2n + 1}. \tag{3.75}$$

We see that the Q function of a Fock state describes a ring with the Bohr–Sommerfeld radius r_n in phase space. This illustrates that Fock states are typical particlelike states containing exactly n energy quanta and showing no wavelike phase dependence. The P function for a Fock state is obtained from Eq. (3.72) by Fourier transformation according to the general relations (3.60) and (3.62). The result

$$P = \frac{1}{n!} \exp\left(\frac{q^2 + p^2}{2}\right) \left[\frac{1}{2}\left(\frac{\partial^2}{\partial q^2} + \frac{\partial^2}{\partial p^2}\right)\right]^n \delta(q)\delta(p) \tag{3.76}$$

indicates clearly that Fock states are indeed nonclassical because their P functions contain derivatives of the two-dimensional delta function $\delta^2(\alpha)$. (The only exception is obviously the vacuum state with $n = 0$.) This example illustrates the mathematical subtleties involved in the P representation.

Let us consider another example, a thermal state, with density operator

$$\hat{\rho} = (1 - e^{-\beta}) \sum_{n=0}^{\infty} |n\rangle \langle n| e^{-n\beta}. \tag{3.77}$$

Here β denotes the ratio $\hbar\omega / k_B T$ of the energy $\hbar\omega$ and the temperature T. (As usual k_B denotes Boltzmann's constant.) To justify the formula (3.77) we recall that in thermal equilibrium the density operator must be diagonal in the energy representation and that photons obey the Bose statistics. We use expression (3.71) to calculate the Q function of the thermal state (3.77)

$$Q(q, p) = (1 - e^{-\beta}) \frac{1}{2\pi} \exp(-|\alpha|^2) \sum_{n=0}^{\infty} \frac{1}{n!} (|\alpha|^2 e^{-\beta})^n$$

$$= \frac{1}{2\pi} (1 - e^{-\beta}) \exp[-|\alpha|^2 (1 - e^{-\beta})] \tag{3.78}$$

$$= \frac{1}{2\pi} (1 - e^{-\beta}) \exp\left[-\frac{1}{2}(q^2 + p^2)(1 - e^{-\beta})\right]. \tag{3.79}$$

The Q function is a Gaussian distribution centered at the origin in phase space. In the limiting case of vanishing temperature, we obtain the Q function $Q_0(q, p)$ of a vacuum, whereas for finite temperature the Gaussian distribution (3.79) is accordingly broader. This difference illustrates the additional fluctuations involved in a thermal state. Using Fourier transformation we obtain from the Q function (3.79) the Wigner function for a thermal state

$$W(q, p) = \frac{1}{\pi} \tanh(\beta/2) \exp[-(q^2 + p^2)\tanh(\beta/2)]. \tag{3.80}$$

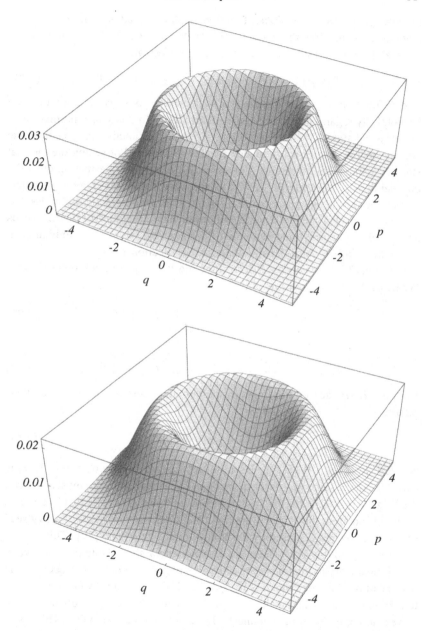

Fig. 3.7. Q functions of a Fock state (top), Eq. (3.72), and of a phase-randomized coherent state (bottom), Eqs. (3.90) and (3.75), with $n = 4$. Although the states are quite different, the Q functions are similar.

Like the Q function, the Wigner function displays the additional thermal fluctuations as well. Note that formula (3.80) also reveals the P function for the thermal density operator (3.77) by Fourier transformation

$$P(q, p) = \frac{1}{\pi}(e^{\beta} - 1) \exp[-(q^2 + p^2)(e^{\beta} - 1)]. \qquad (3.81)$$

The P function of a thermal state is a well-behaved positive function that can be rightfully regarded as a probability distribution. In this sense thermal states are classical states. According to Eq. (3.59) the P function diagonalizes the density operator in terms of coherent states. So instead of seeing the thermal state as a statistical mixture of nonclassical Fock states, we may unravel the thermal density operator (3.77) into a Gaussian distribution of coherent states, that is, into an incoherent mixture of states with well-defined amplitudes and phases. In this way we find a satisfying physical interpretation of thermal states and simultaneously a good example to demonstrate the general ambiguity in unraveling a mixed-state density operator. See Section 1.3.4.

Formula (3.80) reveals the Wigner function $W_n(q, p)$ of Fock states as well. We expand $W(q, p)$ in terms of $e^{-\beta}$

$$W(q, p) = (1 - e^{-\beta}) \sum_{n=0}^{\infty} W_n(q, p) e^{-n\beta} \qquad (3.82)$$

with

$$W_n(q, p) = \frac{(-1)^n}{\pi} \exp(-q^2 - p^2) L_n(2q^2 + 2p^2). \qquad (3.83)$$

Here the $L_n(q)$ denote the Laguerre polynomials, and we have utilized their relation

$$\sum_{n=0}^{\infty} L_n(q) z^n = (1 - z)^{-1} \exp[qz(z - 1)^{-1}]. \qquad (3.84)$$

See Ref. [89], Eq. 10.12(17). Because the thermal density operator (3.77) is expanded in the same way as the expression (3.82), the $W_n(q, p)$ must be indeed the Wigner functions for the Fock states $|n\rangle$. Figure 3.8 shows that in contrast to the Q function, the Wigner function for a Fock state displays a "wavy sea" of rings in the area enclosed by the Bohr–Sommerfeld band [79]. This feature illustrates again that Fock states are clearly nonclassical. Note that the "wavy sea" is necessary to guarantee the orthogonality of the Fock states because the overlap of two Wigner functions $W_n(q, p)$ and $W_{n'}(q, p)$ must vanish according to formula (3.23). The transition to the Q function, however, smoothes out the waves, and only the Bohr–Sommerfeld ring at $(2n + 1)^{1/2}$ remains. This result shows strikingly that the Q function displays far less signature of a quantum state in phase space than does the Wigner function.

We can see this another way. Significantly different quantum states may create similar Q functions. Given a picture of a Q function, it may be difficult

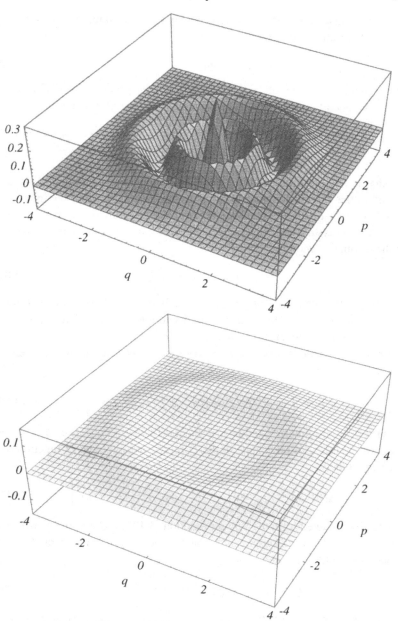

Fig. 3.8. Wigner function of a Fock state with $n = 4$ (top), Eq. (3.83), compared with the Q function (bottom), Eq. (3.72). The plot range for the Q function was set to half of the range for the Wigner function to visualize the differences between the Q and the Wigner representation.

to infer the state behind the picture. Probably the best example to demonstrate this is a Schrödinger-cat state

$$|\psi\rangle \propto |\alpha_0\rangle + |-\alpha_0\rangle \qquad (3.85)$$

(we omit the normalization factor). Using the scalar-product (2.66) of coherent states, we obtain immediately from formula (3.47) the Q function

$$Q(\alpha) \propto \exp(-|\alpha - \alpha_0|^2) + \exp(-|\alpha + \alpha_0|^2)$$
$$+ 2\exp(-|\alpha|^2 - |\alpha_0|^2)\cos[2\text{Im}(\alpha^*\alpha_0)]. \qquad (3.86)$$

All that is left from the beautiful quantum-interference structure clearly displayed in the Wigner function (3.42) is an exponentially small bump proportional to $\exp(-|\alpha_0|^2)$. The more macroscopic the quantum superposition (3.85) is, the smaller is this term. If we neglect the little hump we obtain the Q function of the incoherent mixture

$$\hat{\rho} = \frac{1}{2}(|\alpha_0\rangle\langle\alpha_0| + |-\alpha_0\rangle\langle-\alpha_0|). \qquad (3.87)$$

The Q function cannot clearly discriminate between macroscopic superpositions and statistical mixtures, that is, between the classical *either α_0 or $-\alpha_0$* and the quantum-mechanical α_0 *as well as* $-\alpha_0$.

Another example is a phase-randomized coherent state having the density operator

$$\hat{\rho} = \int_0^{2\pi} |2^{-1/2}r_n\exp(i\phi)\rangle\langle 2^{-1/2}r_n\exp(i\phi)|\frac{d\phi}{2\pi} \qquad (3.88)$$

with the Bohr–Sommerfeld radius r_n defined in (3.75). This state creates almost the same picture as a Fock state in the Q representation. The Q function of a coherent state is

$$Q(q, p) = \frac{1}{2\pi}\exp\left[-\frac{1}{2}(q - q_0)^2 - \frac{1}{2}(p - p_0)^2\right], \qquad (3.89)$$

as is easily obtained from the Wigner function (3.40) or, alternatively, from the scalar product (2.66). Consequently, the Q function of the phase-randomized coherent state (3.88) is given by

$$Q(r) = \frac{1}{2\pi}\exp\left[-\frac{1}{2}(r^2 + r_n^2)\right]I_0(rr_n) \qquad (3.90)$$

using Ref. [225], Eq. 2.5.40.3, to perform the ϕ integration. Here I_0 denotes the zeroth modified Bessel function of the first kind, and the radii r and r_n have been defined in Eqs. (3.74) and (3.75). According to the asymptotic behavior of the Bessel functions for large arguments, expression (3.90) tends rapidly to the approximation

$$Q \sim \frac{1}{(2\pi)^{3/2}(r_n r)^{1/2}} \exp[-(r - r_n)^2] \qquad (3.91)$$

for large quantum numbers. See Ref. [89], Eq. 7.13.1(5). Similar to Fock states, the Q functions of phase-randomized coherent states describe rings in phase space. The only difference is that the rings are broader by a factor of $2^{1/2}$. Yet the states and their physical properties are significantly different. For instance, the photon statistics of phase-randomized coherent states is Poissonian (2.64). (Phase changes do not affect the photon statistics.) In the limit of large intensities the photon distribution (2.64) becomes extremely broad because the variance equals the mean for Poissonian statistics. Fock states, however, always have a precise photon number. Paradoxically, in the region where the photon distributions of phase-randomized coherent states and of Fock states are vastly different, their Q functions are very similar. The Wigner functions, however, differ significantly.

Nevertheless, both the Q function and the P function are mathematically equivalent to any other representation for quantum states, and we are entitled to use this equivalence in tricks to derive theoretical relations. On the other hand, when the Q function is numerically or experimentally given, the retrieval of details hidden in the Q function (but clearly displayed in the Wigner function) takes significant effort in precision. In any case, we must perform a deconvolutionlike procedure that is typically delicate. This fact motivates the conclusion that experimental efforts should be aimed at measuring the Wigner function rather than the Q function to determine the quantum state. The measurement or even the reconstruction of the P function is clearly beyond feasibility, because this mathematical object might be ill-behaved, as we have seen in the case of a Fock state.

3.4 Further reading

Apart from Wigner's Wigner function [290] defined in the phase space of position and momentum, other possible Wigner functions exist for different systems. For instance, spin systems may be described by continuous Wigner functions. See G.S. Agarwal [3]; J.P. Dowling, G.S. Agarwal, and W.P. Schleich [80]; M.O. Scully [248]; L. Cohen and M.O. Scully [57]; M.O. Scully and K. Wódkiewicz [249]; M.O. Scully, H. Walther, and W.P. Schleich [250]; K. Wódkiewicz [295]; and J.C. Várilly and J.M. Gracia-Bondia [277].

The discrete Wigner functions are another intriguing class of quasiprobability distributions for finite-dimensional systems. See the interesting paper by W.K.

Wootters [298] for prime-dimensional state spaces and the extension to odd-dimensional systems by O. Cohendet, Ph. Combe, M. Sirugue, and M. Sirugue-Collin [61], [62]. The Wigner functions for even dimensions are a bit odd, and they, together with the odd-dimensional ones, have been considered in a brief communication [165] and in a detailed paper [170]. Moreover, a discrete Q function has been defined by T. Opatrný, V. Bužek, J. Bajer, and G. Drobný [203] and extended to discrete s-parameterized quasiprobability distributions by T. Opatrný, D.-G. Welsch, and V. Bužek [204].

Also, Wigner functions for angular momentum and phase have been constructed by N. Mukunda [192] and J.P. Bizarro [31]. Wigner functions for photon-number and quantum-optical phase have been given by W.P. Schleich, R.J. Horowicz, and S. Varro [242]; J.A. Vaccaro and D.T. Pegg [271]; A. Lukš and V. Peřinová [183]; and, finally, in a communication [272] and in a comprehensive paper [274] by J.A. Vaccaro.

4

Simple optical instruments

4.1 Beam splitter

Compared with other fundamental experiments in physics, optical tests of quantum mechanics are often distinguished by their simplicity. Most quantum-optical experiments do not require a whole industry – an optical table of equipment, a few people, and yet the right question to ask are often sufficient. "Research is to see what everybody has seen and to think what nobody has thought" [125]. A simple optical beam splitter, for instance, is already a nice device to demonstrate the quantum nature of light. Quite a number of puzzling quantum effects have been seen by splitting or recombining photons with a small cube of glass. Additionally, the beam splitter serves as a theoretical paradigm for other linear optical devices. Interferometers, semitransparent mirrors, dielectric interfaces, wave-guide couplers, and polarizers are all described sufficiently well by a simple beam-splitter model. This model also accounts for the effect of absorption, mode mismatch, and other linear losses. The quantum effects of almost all passive optical devices can be understood assuming appropriate beam-splitter models. (It's all done with mirrors.)

4.1.1 Heisenberg picture

An ideal beam splitter is a reversible, lossless device in which two incident beams may interfere to produce two emerging beams. For instance, a dielectric interface inside a cube or plate of glass splits a light beam into two. We may reverse this situation by sending the two beams back to the cube, where they interfere constructively to restore the original beam. However, if we change the phases of the two beams, their mutual interference generates two emerging beams in general. So four beams might be involved, two incident and two outgoing light modes, and the splitting of just one beam is a special case. Most

67

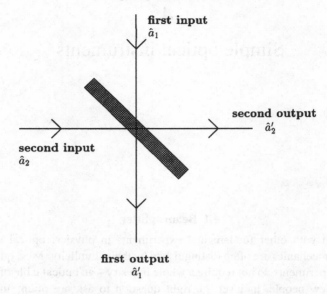

Fig. 4.1. Diagram of a lossless beam splitter. Two incident spatial–temporal light modes (with the annihilation operators $\hat{a}_1$ and $\hat{a}_2$) interfere optically to produce two emerging beams (with the annihilation operators $\hat{a}'_1$ and $\hat{a}'_2$).

polarizing beam splitters use anisotropic media (for instance, calcite) to split the two polarizations of the incident field into two spatially separate beams. Wave-guide couplers apply the optical tunneling effect for mixing the light fields traveling in two fibers. All these simple optical instruments act like beam splitters. To make our theoretical beam-splitter model as general as possible, let us describe the device as a *four-port*, that is, as a black box with two input and two output ports having certain properties.

What happens if two coherent light beams with complex amplitudes α_1 and α_2 interfere? In classical optics the amplitudes are simply superimposed according to the linear transformation

$$\begin{pmatrix} \alpha'_1 \\ \alpha'_2 \end{pmatrix} = \underline{B} \begin{pmatrix} \alpha_1 \\ \alpha_2 \end{pmatrix} \tag{4.1}$$

described by the matrix

$$\underline{B} = \begin{pmatrix} B_{11} & B_{12} \\ B_{21} & B_{22} \end{pmatrix}. \tag{4.2}$$

In quantum optics the complex amplitudes α_1 and α_2 correspond to the annihilation operators $\hat{a}_1$ and $\hat{a}_2$ of the incident fields, and the emerging beams are characterized by the operators $\hat{a}'_1$ and $\hat{a}'_2$. We assume that the linear interference (4.1) is also valid for quantum fields, that is,

$$\begin{pmatrix} \hat{a}'_1 \\ \hat{a}'_2 \end{pmatrix} = \underline{B} \begin{pmatrix} \hat{a}_1 \\ \hat{a}_2 \end{pmatrix}. \tag{4.3}$$

This simple model describes very well most passive and lossless devices in which two input beams interfere to produce two outgoing fields. It is quite typical that classical laws rule interference phenomena in the classical as well as in the quantum domain. Usually quantum effects change only the visibility of the interference but not the phenomenon itself. We also note that our model (4.3) may be justified using more sophisticated theories [281] of the quantum propagation of light in dielectric media.

Because the model (4.3) describes the change of the operators $\hat{a}_1$ and $\hat{a}_2$ after beam splitting, it corresponds to the Heisenberg picture of the process. We determine easily the general properties of the beam-splitter matrix $\underline{B}$ using this picture. Assuming that the incoming and the emerging beams are both independent bosonic modes, their annihilation operators must satisfy

$$[\hat{a}'_\nu, \hat{a}'^\dagger_\mu] = [\hat{a}_\nu, \hat{a}^\dagger_\mu] = \delta_{\nu\mu}, \tag{4.4}$$

$$[\hat{a}'_\nu, \hat{a}'_\mu] = [\hat{a}_\nu, \hat{a}_\mu] = 0. \tag{4.5}$$

Consequently, the beam-splitter matrix $\underline{B}$ must obey

$$|B_{11}|^2 + |B_{12}|^2 = |B_{21}|^2 + |B_{22}|^2 = 1, \tag{4.6}$$

$$B_{11}B^*_{21} + B_{12}B^*_{22} = 0 \tag{4.7}$$

or, in other words, $\underline{B}$ is unitary

$$\underline{B}^{-1} = \underline{B}^\dagger. \tag{4.8}$$

This condition reflects the fact that a lossless beam splitter conserves energy and that the total intensity $\hat{a}^\dagger_1\hat{a}_1 + \hat{a}^\dagger_2\hat{a}_2$ is thus an invariant quantity. Apart from the unitarity (4.8) the beam-splitter coefficients are free parameters that depend on the particular experimental situation. They comprise the material properties of an employed interferometer, a linear coupler, or a polarizer. As is done in our general simplified model for light beams, the electromagnetic oscillator (see Section 2.1), we have reduced the wealth of the classical optics involved to a few material parameters. Additionally, we have assumed that we are dealing with only four well-defined optical modes. Our simplified model allows us to focus on the essential quantum properties of four-ports and to be general.

To proceed further we recall the well-known mathematical structure of two-dimensional unitary matrices. Any unitary $\underline{B}$ can be represented as the matrix product

$$\underline{B} = e^{i\Lambda/2} \begin{pmatrix} e^{i\Psi/2} & 0 \\ 0 & e^{-i\Psi/2} \end{pmatrix} \begin{pmatrix} \cos(\Theta/2) & \sin(\Theta/2) \\ -\sin(\Theta/2) & \cos(\Theta/2) \end{pmatrix} \begin{pmatrix} e^{i\Phi/2} & 0 \\ 0 & e^{-i\Phi/2} \end{pmatrix}$$

(4.9)

with the real numbers Λ, Θ, Ψ, and Φ or, expressed explicitly,

$$\underline{B} = e^{i\Lambda/2} \begin{pmatrix} \cos(\Theta/2)e^{i(\Psi+\Phi)/2} & \sin(\Theta/2)e^{i(\Psi-\Phi)/2} \\ -\sin(\Theta/2)e^{i(-\Psi+\Phi)/2} & \cos(\Theta/2)e^{i(-\Psi-\Phi)/2} \end{pmatrix}. \tag{4.10}$$

Any four-port can be considered as acting in three steps. First the phases of the incident modes are changed, then the amplitudes are mixed (rotated), and finally the phases are changed again. In many cases we can incorporate the phase shifts in the definition of the reference phases of the incoming or emerging beams. The rotation of the mode operators, however, remains the key feature of four-ports. In most later calculations concerning beam splitters, we will consider only real rotation matrices $\underline{B}$. In this case we may represent $\underline{B}$ in terms of the *transmissivity* τ and the *reflectivity* ϱ as

$$\underline{B} = \begin{pmatrix} \tau & -\varrho \\ \varrho & \tau \end{pmatrix}. \tag{4.11}$$

In this notation τ equals $\cos(\Theta/2)$, ϱ means $-\sin(\Theta/2)$, and the relation

$$\tau^2 + \varrho^2 = 1 \tag{4.12}$$

accounts for the energy conservation of the lossless four-port.

Note that we have silently smuggled in one essential quantum feature of beam splitters: A beam splitter is a four-port not only in the case of two incoming fields interfering to produce two emerging beams; a beam splitter is always a four-port. Even if only one beam is split into two, if literally nothing behind the semitransparent mirror is interfering with the incident field, quantum-mechanically this nothing means a vacuum state. The sheer possibility that the second mode behind the mirror might be excited makes a difference. The vacuum fluctuations carried by the empty mode (and entering the apparatus via the so-called *unused port* of the beam splitter) do cause physical effects. In particular, we will see in Section 6.2 that this picture of the vacuum fluctuations behind the mirror is useful for understanding a fundamental quantum-optical experiment. From a mere formal point of view the explanation of this intriguing quantum feature is elementary in the Heisenberg picture. Suppose that only one beam described by the annihilation operator $\hat{a}_1$ would split into two modes corresponding to the operators $\hat{a}'_1$ and $\hat{a}'_2$ according to the linear transformations $\hat{a}'_1 = B_1\hat{a}_1$ and $\hat{a}'_2 = B_2\hat{a}_1$. Because the commutator $[\hat{a}'_1, \hat{a}'^\dagger_2]$ gives $B_1 B_2^*$ and not zero, the emerging modes cannot be independent quantum systems. The introduction of the second beam, $\hat{a}_2$, however, and the unitarity of the beam-splitter matrix guarantee that the outgoing fields can be considered independent bosonic modes.

4.1.2 Schrödinger picture

In the Heisenberg picture the state of the incident beams is invariant, whereas the mode operators $\hat{a}_1$ and $\hat{a}_2$ are changed according to the linear transformation (4.3). In the Schrödinger picture we encounter the opposite situation – the operators are invariant, whereas the states are changed. The standard way of deducing the behavior in the Schrödinger representation from the Heisenberg picture (and vice versa) is to find an evolution operator. This operator, here denoted by $\hat{B}$, performs the transformation (4.3) of the observables in the Heisenberg representation

$$\begin{pmatrix} \hat{a}_1' \\ \hat{a}_2' \end{pmatrix} = \hat{B} \begin{pmatrix} \hat{a}_1 \\ \hat{a}_2 \end{pmatrix} \hat{B}^\dagger. \tag{4.13}$$

In the Schrödinger picture the density operator $\hat{\rho}$ is changed accordingly

$$\hat{\rho}' = \hat{B}^\dagger \hat{\rho} \hat{B}. \tag{4.14}$$

Both formulas (4.13) and (4.14) are designed to produce identical expectation values $\mathrm{tr}\{\hat{\rho} F(\hat{a}_1, \hat{a}_2)\}$, and hence both pictures are considered physically equivalent. In the case of a pure state, the state vector $|\psi\rangle$ is transformed as

$$|\psi\rangle' = \hat{B}^\dagger |\psi\rangle. \tag{4.15}$$

A convenient trick for finding the desired operator $\hat{B}$ of the beam splitting is to employ the Jordan–Schwinger representation [132], [247] of an angular momentum in terms of two bosonic operators, one for each mode,

$$\hat{L}_0 = \frac{1}{2} \left(\hat{a}_1^\dagger \hat{a}_1 + \hat{a}_2^\dagger \hat{a}_2 \right),$$

$$\hat{L}_1 = \frac{1}{2} \left(\hat{a}_1^\dagger \hat{a}_2 + \hat{a}_2^\dagger \hat{a}_1 \right),$$

$$\hat{L}_2 = \frac{1}{2\mathrm{i}} \left(\hat{a}_1^\dagger \hat{a}_2 - \hat{a}_2^\dagger \hat{a}_1 \right),$$

$$\hat{L}_3 = \frac{1}{2} \left(\hat{a}_1^\dagger \hat{a}_1 - \hat{a}_2^\dagger \hat{a}_2 \right). \tag{4.16}$$

An easy exercise verifies that $\hat{L}_1$, $\hat{L}_2$, and $\hat{L}_3$ obey the commutation relations of angular-momentum components. The operator $\hat{L}_0$ commutes with all other $\hat{L}_k$, and $\hat{L}_0(\hat{L}_0 + 1) = \hat{L}_1^2 + \hat{L}_2^2 + \hat{L}_3^2$ represents the total angular momentum $\hat{L}^2$. These properties show that the Jordan–Schwinger operators (4.16) behave

indeed like angular-momentum components. Note that a simple physical interpretation exists for this analogy if the two involved modes are the two possible
polarizations of a light beam. In this case the Jordan–Schwinger operators are
simply the quantized Stokes parameters [38] describing the degree of polarization or, in other words, the spin properties of light. (In fact, the operators (4.16)
are deeply related to the quantum theory of partially polarized or unpolarized
light [150].)

How do the Jordan–Schwinger components transform the mode operators?
We note that, for a constant Λ,

$$\exp(-\mathrm{i}\Lambda\hat{L}_0)\begin{pmatrix}\hat{a}_1\\\hat{a}_2\end{pmatrix}\exp(\mathrm{i}\Lambda\hat{L}_0) = \exp(\mathrm{i}\Lambda/2)\begin{pmatrix}\hat{a}_1\\\hat{a}_2\end{pmatrix} \qquad (4.17)$$

and, for a constant Φ,

$$\exp(-\mathrm{i}\Phi\hat{L}_3)\begin{pmatrix}\hat{a}_1\\\hat{a}_2\end{pmatrix}\exp(\mathrm{i}\Phi\hat{L}_3) = \begin{pmatrix}\exp(\mathrm{i}\Phi/2)\hat{a}_1\\\exp(-\mathrm{i}\Phi/2)\hat{a}_2\end{pmatrix}, \qquad (4.18)$$

as is easily seen from the basic property (2.7) of the phase-shifting operator $\hat{U}$
defined in Eq. (2.6). We also note that, for another constant Θ,

$$\exp(-\mathrm{i}\Theta\hat{L}_2)\begin{pmatrix}\hat{a}_1\\\hat{a}_2\end{pmatrix}\exp(\mathrm{i}\Theta\hat{L}_2) = \begin{pmatrix}\cos(\Theta/2) & \sin(\Theta/2)\\-\sin(\Theta/2) & \cos(\Theta/2)\end{pmatrix}\begin{pmatrix}\hat{a}_1\\\hat{a}_2\end{pmatrix}. \qquad (4.19)$$

The easiest way of verifying this formula is to show that both sides obey the same
differential equation with respect to Θ. This property is seen by differentiating
and considering the commutation relations of $\hat{L}_2$ and the annihilation operators
$\hat{a}_1$ and $\hat{a}_2$. Because both sides of Eq. (4.19) agree obviously for $\Theta = 0$, they
must be indeed identical.

Now we have everything on hand to see from the representation (4.9) of the
$\underline{B}$ matrix that the evolution operator $\hat{B}$ is given by

$$\hat{B} = \exp(-\mathrm{i}\Phi\hat{L}_3)\exp(-\mathrm{i}\Theta\hat{L}_2)\exp(-\mathrm{i}\Psi\hat{L}_3)\exp(-\mathrm{i}\Lambda\hat{L}_0) \qquad (4.20)$$

for the constants Φ, Θ, Ψ, and Λ. To flesh out this formal construction, let
us consider the effect of a real beam-splitter transformation (4.11) on the
Schrödinger wave function for the quadratures q_1 and q_2. As pointed out in the
previous section, the case of a real $\underline{B}$ matrix contains the physical essence of
any beam-splitter transformation. According to rule (4.15) and representation
(4.20), the two-mode wave function

$$\psi(q_1, q_2) \equiv \langle q_1|\langle q_2|\psi\rangle \qquad (4.21)$$

of the incident beams is changed to

$$\psi'(q_1, q_2) = \langle q_1|\langle q_2|\exp(\mathrm{i}\Theta\hat{L}_2)|\psi\rangle \qquad (4.22)$$

to produce the wave function $\psi'(q_1, q_2)$ of the emerging beams. We express $\hat{L}_2$ defined in (4.16) in terms of the quadratures using Eq. (2.13)

$$\hat{L}_2 = \frac{1}{2}(\hat{q}_1 \hat{p}_2 - \hat{q}_2 \hat{p}_1) \tag{4.23}$$

and obtain in the Schrödinger representation

$$\psi'(q_1, q_2) = \exp\left[\frac{\Theta}{2}\left(q_1 \frac{\partial}{\partial q_2} - q_2 \frac{\partial}{\partial q_1}\right)\right] \psi(q_1, q_2) \tag{4.24}$$

$$= \psi(q_1', q_2') \tag{4.25}$$

with

$$\begin{pmatrix} q_1' \\ q_2' \end{pmatrix} = \begin{pmatrix} \cos(\Theta/2) & -\sin(\Theta/2) \\ \sin(\Theta/2) & \cos(\Theta/2) \end{pmatrix} \begin{pmatrix} q_1 \\ q_2 \end{pmatrix}. \tag{4.26}$$

In the last step we have used the fact that formula (4.25) obeys the same differential equation with respect to Θ as the expression (4.24) does. In fact,

$$\frac{\partial \psi'}{\partial \Theta} = \left(\frac{\partial \psi'}{\partial q_1} \frac{\partial q_1}{\partial q_1'} + \frac{\partial \psi'}{\partial q_2} \frac{\partial q_2}{\partial q_1'}\right)\frac{\partial q_1'}{\partial \Theta} + \left(\frac{\partial \psi'}{\partial q_1}\frac{\partial q_1}{\partial q_2'} + \frac{\partial \psi'}{\partial q_2}\frac{\partial q_2}{\partial q_2'}\right)\frac{\partial q_2'}{\partial \Theta}$$

$$= \frac{1}{2}\left(q_1 \frac{\partial}{\partial q_2} - q_2 \frac{\partial}{\partial q_1}\right)\psi'(q_1, q_2). \tag{4.27}$$

Because the formulas (4.24) and (4.25) agree for $\Theta = 0$, they must be identical for all Θ values. The wave function is simply rotated. In other words, a beam splitter transforms the quadrature wave function in a classical way:

$$\psi'(q_1, q_2) = \psi(\tau q_1 + \varrho q_2, \varrho q_1 - \tau q_2), \tag{4.28}$$

with $\tau = \cos(\Theta/2)$ and $\varrho = -\sin(\Theta/2)$. In many cases the rotation of the wave function provides us with a simple intuitive picture for the quantum effects of these devices [155]. For example, if two beams of equally squeezed vacuums interfere according to the real transformation (4.11), they leave the apparatus unchanged simply because their total wave function

$$\psi(q_1, q_2) = e^{\zeta} \pi^{-1/2} \exp\left[-\frac{1}{2} e^{2\zeta}\left(q_1^2 + q_2^2\right)\right] \tag{4.29}$$

is isotropic. (We have used Eqs. (2.33) and (2.79) for the wave functions of the squeezed vacuums.) On the other hand, if two oppositely squeezed vacuums

$$\psi_1(q_1) = e^{\zeta/2} \pi^{-1/4} \exp\left(-\frac{1}{2} e^{2\zeta} q_1^2\right), \tag{4.30}$$

$$\psi_2(q_2) = e^{-\zeta/2} \pi^{-1/4} \exp\left(-\frac{1}{2} e^{-2\zeta} q_2^2\right) \tag{4.31}$$

interfere at a real $50:50$ beam splitter (where $\tau = \varrho = 2^{-1/2}$), they produce a *two-mode squeezed vacuum*

$$\psi'(q_1, q_2) = \pi^{-1/2} \exp\left[-\frac{1}{4}e^{2\varsigma}(q_1 + q_2)^2 - \frac{1}{4}e^{-2\varsigma}(q_1 - q_2)^2\right]. \quad (4.32)$$

This wave function describes the strongest entangled state possible (with given mean energy) [20], [21], and it provides a physical realization [207], [231], [232] of the original Einstein–Podolsky–Rosen state [84]. So the squeezing directions determine whether the two interfering modes produce a perfectly disentangled or a maximally correlated pair of beams [35]. Of course, we may also invert the process and disentangle the modes of a two-mode squeezed vacuum by optical interference to gain two independent beams of squeezed light. This is, by the way, related to the standard method of making single-mode squeezing in type II parametric down-conversion [174]. Here, photons of the pump beam are converted into pairs of orthogonally polarized photons called *signal* and *idler*. Rotating the polarization axis by $45°$ produces single-mode squeezed light fields as linear combinations of the generated signal and idler photons. This and many other quantum properties of polarizers and beam splitters can be easily explained with the simple picture of rotating wave functions [155].

How is the Wigner function transformed under the action of a beam splitter? Clearly, two modes are involved, and so we must describe the quantum state of the incident beams by a two-mode Wigner function extending Wigner's formula (3.17) to two degrees of freedom

$$W(q_1, p_1, q_2, p_2) = \frac{1}{(2\pi)^2} \int_{-\infty}^{+\infty} \int_{-\infty}^{+\infty} \exp[i(p_1 x_1 + p_2 x_2)]$$
$$\times \left\langle q_1 - \frac{x_1}{2}, q_2 - \frac{x_2}{2} \middle| \hat{\rho} \middle| q_1 + \frac{x_2}{2}, q_2 + \frac{x_2}{2} \right\rangle$$
$$\times dx_1 dx_2. \quad (4.33)$$

For understanding the action of a beam splitter in the Wigner representation, let us focus on the mode rotation (4.11) only. (The effect of the phase shifts has been considered already in the definition (3.1) of the Wigner function.) Because quadrature wave functions are simply rotated, the Wigner function is rotated as well,

$$W'(q_1, p_1, q_2, p_2) = W(\tau q_1 + \varrho q_2, \tau p_1 + \varrho p_2, \varrho q_1 - \tau q_2, \varrho p_1 - \tau p_2), \quad (4.34)$$

as is easily seen using Wigner's formula extended to the two-mode case (4.33). The points of the Wigner function move exactly like classical quantities distributed according to $W(q_1, p_1, q_2, p_2)$. In this respect the Wigner function behaves like a classical phase-space density. Note, however, that this feature of

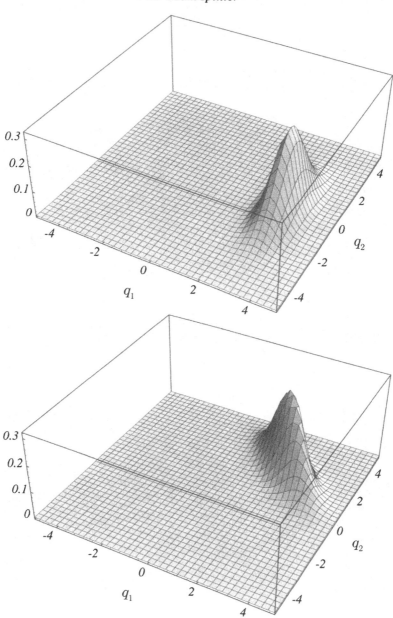

Fig. 4.2. Optical interference of two squeezed states with opposite squeezing parameters. Plot of the total position-quadrature distribution $|\psi(q_1, q_2)|^2$ before optical interference (top) and after (bottom). The first incident beam has a squeezing parameter ζ of 0.5 and a position displacement q_0 of 3. The second beam is a squeezed vacuum with $\zeta = -0.5$. [See Eq. (2.85) for the wave functions.] The rotated position-quadrature distribution after the interference visualizes the quantum entanglement of the emerging beams.

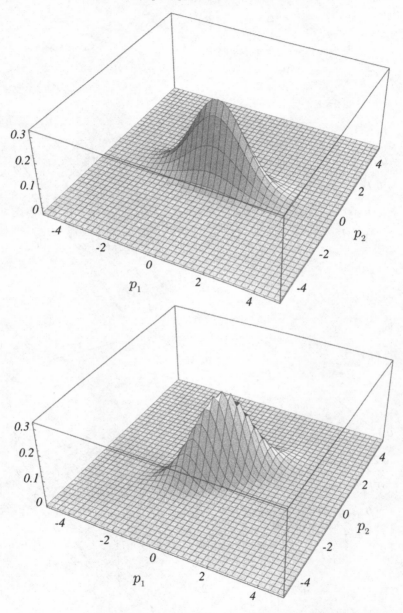

Fig. 4.3. Optical interference of two squeezed states with opposite squeezing param-
eters. Plot of the total momentum-quadrature distribution $|\psi(p_1, p_2)|^2$ before optical
interference (top) and after (bottom). The first incident beam has a squeezing parameter
ζ of 0.5 and no momentum displacement p_0. The second beam is a squeezed vacuum
with $\zeta = -0.5$. [The wave functions are easily obtained by a Fourier transformation of
Eq. (2.85).] The rotated momentum-quadrature distribution after the interference shows
strikingly the high quadrature correlations of the emerging beams.

the Wigner representation is restricted to linear transformations, as was noticed early by Moyal [191]. Nonlinear effects cause a complicated reshaping of the Wigner function in general. Formula (4.34), though a simple mathematical relation, is the key for understanding the behavior of beam splitters whenever phase-space quantities are concerned. In particular, the formula will provide us with an elegant description of eight-port-homodyne detection, as we will see in Chapter 6. We also notice immediately from the expression (3.40) of the Wigner function for coherent states that the optical interference of coherent light produces just coherent light. To be more precise, if the incident beams are in the coherent states $|\alpha_1\rangle$ and $|\alpha_2\rangle$, the emerging fields are in the coherent states $|\alpha_1'\rangle$ and $|\alpha_2'\rangle$, with α_1' and α_2' given by the classical interference law (4.1). This is consistent with our motivation for the quantum theory of a beam splitter.

4.1.3 Fock representation and wave–particle dualism

We have considered beam splitting in the Heisenberg and Schrödinger picture and using Wigner functions. What happens in the photon-number representation? How are photons distributed from an incident light beam to the emerging field modes? We express the density operator in the Fock basis

$$\hat{\rho} = \sum_{n_1,n_2=0}^{\infty} \sum_{m_1,m_2=0}^{\infty} \rho(n_1, n_2; m_1, m_2)|n_1, n_2\rangle\langle m_1, m_2| \qquad (4.35)$$

with the density matrix

$$\rho(n_1, n_2; m_1, m_2) = \langle n_1, n_2|\hat{\rho}|m_1, m_2\rangle. \qquad (4.36)$$

Here $|n_1, n_2\rangle$ denotes a two-mode Fock state $|n_1\rangle \otimes |n_2\rangle$ generated from the vacuum state $|0, 0\rangle$ by the formal procedure

$$|n_1, n_2\rangle = (n_1!n_2!)^{-1/2}\hat{a}_1^{\dagger n_1}\hat{a}_2^{\dagger n_2}|0, 0\rangle \qquad (4.37)$$

according to relation (2.35). In the Schrödinger picture the observables are invariant, whereas the state is changed by the transformation operator $\hat{B}$. See Eq. (4.15). To find the transformed density operator

$$\hat{\rho}' = \sum_{n_1,n_2=0}^{\infty} \sum_{m_1,m_2=0}^{\infty} \rho(n_1, n_2; m_1, m_2)\hat{B}^{\dagger}|n_1, n_2\rangle\langle m_1, m_2|\hat{B} \qquad (4.38)$$

describing the quantum state of the emerging beams, we need to consider the beam-splitting of the Fock states

$$\hat{B}^{\dagger}|n_1, n_2\rangle = (n_1!n_2!)^{-1/2}\hat{B}^{\dagger}\hat{a}_1^{\dagger n_1}\hat{a}_2^{\dagger n_2}\hat{B}|0, 0\rangle. \qquad (4.39)$$

Note that this equation involves a subtlety. We have replaced $|0, 0\rangle$ by $\hat{B}|0, 0\rangle$, using the fact that an incident two-mode vacuum produces just vacuum, because the beam splitter conserves energy. We express in Eq. (4.39) the "old" mode operators $\hat{a}_1$ and $\hat{a}_2$ in terms of the "new" operators $\hat{a}'_1$ and $\hat{a}'_2$. We simply invert the basic transformation (4.3),

$$\begin{pmatrix} \hat{a}_1 \\ \hat{a}_2 \end{pmatrix} = \underline{B}^\dagger \begin{pmatrix} \hat{a}'_1 \\ \hat{a}'_2 \end{pmatrix} = \begin{pmatrix} B_{11}^* & B_{21}^* \\ B_{12}^* & B_{22}^* \end{pmatrix} \begin{pmatrix} \hat{a}'_1 \\ \hat{a}'_2 \end{pmatrix}, \qquad (4.40)$$

which is easily done because the beam-splitter matrix $\underline{B}$ is unitary. Consequently, we get via the transformation (4.13) in the Heisenberg picture

$$\hat{B}^\dagger |n_1, n_2\rangle = (n_1! n_2!)^{-1/2} \left(B_{11}\hat{a}_1^\dagger + B_{21}\hat{a}_2^\dagger \right)^{n_1} \left(B_{12}\hat{a}_1^\dagger + B_{22}\hat{a}_2^\dagger \right)^{n_2} |0, 0\rangle. \quad (4.41)$$

So, in essence, we have simply replaced the "old" mode operators by the "new" mode operators in definition (4.37) to find an expression for the optically mixed Fock states. We expand the products in Eq. (4.41), use definition (4.37) again, and obtain the lengthy formula

$$\hat{B}^\dagger |n_1, n_2\rangle = (n_1! n_2!)^{-1/2} \sum_{k_1, k_2 = 0}^{n_1, n_2} \binom{n_1}{k_1} \binom{n_2}{k_2} (B_{11})^{k_1} (B_{21})^{n_1 - k_1}$$

$$\times (B_{12})^{k_2} (B_{22})^{n_2 - k_2} ((k_1 + k_2)!(n_1 + n_2 - k_1 - k_2)!)^{1/2}$$

$$\times |k_1 + k_2, n_1 + n_2 - k_1 - k_2\rangle. \qquad (4.42)$$

Quite in contrast to the Schrödinger representation, the beam-splitter transformation in the Fock basis is rather involved. A few instructive special cases of the general formula (4.42) are, however, worth considering in detail.

What happens if precisely n photons are "split"? In this case the first beam is in the Fock state $|n\rangle$, whereas the second mode is a vacuum. We obtain from formula (4.42), assuming a real beam-splitter matrix,

$$\hat{B}^\dagger |n, 0\rangle = \sum_{k=0}^{n} \binom{n}{k}^{1/2} \tau^k \varrho^{n-k} |k, n - k\rangle. \qquad (4.43)$$

Of course, photons are not split to fractions. Rather, they are distributed by the beam splitter to the two emerging fields. The probability of finding k photons in beam one (and consequently $n - k$ photons in beam two) is given by $\binom{n}{k}(\tau^2)^k (\varrho^2)^{n-k}$. This rule is precisely the statistical law for distributing n distinguishable classical particles to two channels, the law of a photon lottery. What happens there? The probability that an individual particle goes to the first channel should be τ^2, and the probability for a move to the second channel should be ϱ^2. [The sum $\tau^2 + \varrho^2$ gives unity, as is guaranteed by energy conservation (4.12)]. Consequently, the probability of finding k selected particles in beam one and $n - k$ individuals in beam two is given by the product $(\tau^2)^k (\varrho^2)^{n-k}$. For us, all photons are equal, and so we cannot discriminate between individual particles. We observe the probability of having *any* k particles in beam one and any $n - k$ in beam two and obtain in this way a statistical enhancement given by the number of all combinations, that is, by the binomial coefficient $\binom{n}{k}$.

Splitting precisely n photons shows strikingly the particle nature of light, as has been demonstrated experimentally [42]. Photons are indivisible energy quanta, and so the beam splitter faces no other choice than distributing them statistically. On the other hand, what happens if two photons interfere optically? To be precise, suppose that each of both incident beams carries exactly one photon. What would we expect? Individual photons cannot have definite optical phases. However, if they are forced to interfere either constructively or destructively, they might consider both options as equally likely and emerge in pairs in beam one or in beam two. That this is indeed true is easily seen from formula (4.42) assuming a real $50:50$ beam splitter with $\tau = \varrho = 2^{-1/2}$. In fact, we obtain

$$\hat{B}^{\dagger}|1, 1\rangle = 2^{-1/2}(|0, 2\rangle - |2, 0\rangle). \tag{4.44}$$

Photons emerge only in pairs. To be more precise, photons emerge in the quantum superposition state of being both together in one emerging beam or in the other. Consequently, if one photon is detected in one beam, then no photon should be in the other. The count correlation of the two emerging beams should exhibit a pronounced minimum that was experimentally demonstrated by Hong, Ou, and Mandel [118] for the first time. We have seen that the splitting or the interference of single photons is a wonderful way of demonstrating the fundamental wave–particle dualism of light.

4.1.4 Beam-splitter model of absorption

What happens if a light beam has been partially absorbed by a linear medium? We would expect that the complex wave amplitude α is reduced by a certain factor that is independent of the absorbed field (because the medium response is linear). To be more precise, we assume that any coherent state $|\alpha\rangle$ is reduced in intensity by the factor η (with $0 < \eta \le 1$) so that after partial absorption the field is in the coherent state $|\eta^{1/2}\alpha\rangle$. What happens with other initial states? To find a motivation, we express the density operator $\hat{\rho}$ in terms of the P function according to Eq. (3.59). In this form the quantum state appears as a distribution $P(\alpha)$ of coherent states $|\alpha\rangle$ (although $P(\alpha)$ might be ill-behaved). We assume that we can apply the absorption rule for coherent states to each individual term of the representation (3.59) and obtain for the density operator $\hat{\rho}'$ of the partially absorbed mode

$$\hat{\rho}' = \int_{-\infty}^{+\infty} \int_{-\infty}^{+\infty} P(\alpha)|\eta^{1/2}\alpha\rangle\langle\eta^{1/2}\alpha|\, dq\, dp \qquad (4.45)$$

with $\alpha = 2^{-1/2}(q + ip)$. We change variables

$$\hat{\rho}' = \int_{-\infty}^{+\infty} \int_{-\infty}^{+\infty} \eta^{-1} P(\eta^{-1/2}\alpha)|\alpha\rangle\langle\alpha|\, dq\, dp \qquad (4.46)$$

and see that the absorber is simply scaling the argument of the P function. Consequently, the Fourier-transformed P distribution (the characteristic function with parameter $s = +1$) is scaled accordingly

$$\tilde{W}'(u, v; +1) = \tilde{W}(\eta^{1/2}u, \eta^{1/2}v; +1). \qquad (4.47)$$

Using this relation and the definition (3.60) of s-parameterized characteristic functions, we obtain the general transformation law for absorption

$$\tilde{W}'(u, v; s) = \tilde{W}(\eta^{1/2}u, \eta^{1/2}v; s') \qquad (4.48)$$

with the changed parameter

$$s' = \frac{1}{\eta}(s + \eta - 1). \qquad (4.49)$$

Consequently, absorbers scale quasiprobability distributions

$$W'(q, p; s) = \eta^{-1} W(\eta^{-1/2}q, \eta^{-1/2}p; s') \qquad (4.50)$$

and modify their parameters according to the simple rule (4.49). In particular, the Wigner function is transformed into an s'-parameterized distribution with $s' = -(\eta^{-1} - 1)$, and instead of the Q function we obtain a distribution with the parameter $-(2\eta^{-1} - 1)$.

To see a simple intuitive picture to visualize the quantum effects of linear absorption, we express the transformed Wigner function $W'(q, p)$ using the general relation (3.65) of s-parameterized quasiprobability distributions

$$\tilde{W}'(q, p) = \frac{1}{(1 - \eta)\pi} \int_{-\infty}^{+\infty} \int_{-\infty}^{+\infty} W(q', p')$$

$$\times \exp\left[-\frac{\eta}{1 - \eta}((q' - \eta^{-1/2}q)^2 + (p' - \eta^{-1/2}p)^2)\right] dq'\, dp'. \qquad (4.51)$$

We replace the integration variables q' and p' by $\eta^{1/2}q - (1 - \eta)^{1/2}q_2$ and $\eta^{1/2}p - (1 - \eta)^{1/2}p_2$, respectively, and obtain

$$W'(q, p) = \int_{-\infty}^{+\infty} \int_{-\infty}^{+\infty} W[\eta^{1/2}q - (1 - \eta)^{1/2}q_2, \eta^{1/2}p - (1 - \eta)^{1/2}p_2]$$

$$\times W_0[(1 - \eta)^{1/2}q + \eta^{1/2}q_2, (1 - \eta)^{1/2}p + \eta^{1/2}p_2]\, dq_2\, dp_2. \qquad (4.52)$$

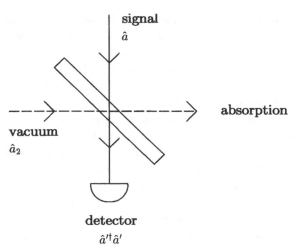

Fig. 4.4. A fictitious beam splitter is a convenient model for describing linear absorption or, equivalently, detection losses. The signal is attenuated and, simultaneously, contaminated by the vacuum fluctuations entering the second input port of the fictitious beam splitter.

Here $W_0(q, p)$ abbreviates the Gaussian $\pi^{-1} \exp(-q^2 - p^2)$. This Gaussian is precisely the Wigner function for a vacuum. See Eq. (3.32). We recall that beam splitters rotate the Wigner function and find immediately a simple physical interpretation for formula (4.52): The absorber acts like a fictitious beam splitter. When light is damped it can be imagined as being split into a transmitted part and an absorbed part. On the other hand, we know from the fluctuation–dissipation theorem (see for instance Ref. [187], Section 17.2) that losses are always accompanied by fluctuations. At least the vacuum fluctuations of the absorbing medium must be taken into account. In our simple absorber model depicted in Fig. 4.4, these fluctuations come into play via the unused second port of the fictitious beam splitter.

We translate the Wigner representation (4.52) to the Heisenberg picture for our fictitious beam splitter and see that there the annihilation operator $\hat{a}$ of the partially absorbed mode is transformed according to

$$\hat{a}' = \eta^{1/2}\hat{a} + (1 - \eta)^{1/2}\hat{a}_2. \tag{4.53}$$

Here $\hat{a}_2$ denotes the formal mode operator of the vacuum field entering the unused port in our model. In this picture we easily understand why the fluctuation mode is unavoidable. The second term $(1 - \eta)^{1/2}\hat{a}_2$ in (4.53) is necessary to guarantee that the damped field remains a proper bosonic mode, because otherwise the commutation relation $[\hat{a}', \hat{a}'^{\dagger}] = 1$ would be violated. We also find a simple physical interpretation of the way (4.50) in which absorption changes quasiprobability distributions. The rescaling corresponds to the linear loss in intensity, whereas the s-parameter change mirrors the introduced extra fluctuations. In addition, our model shows why losses and other inefficiencies should be avoided in experiments with nonclassical light. Absorption smoothes the Wigner function of the considered quantum state and leads ultimately to a loss of nonclassical behavior [130]. Note that our simple beam-splitter model is also correct for thermal or phase-sensitive damping [155]. In these cases the fictitious field entering the second port is in a thermal or squeezed state. The whole procedure of doubling the system by introducing a formal fluctuation mode is also known as the thermo-field technique [93], [270]. Finally, we note that the model describes detection efficiencies [303] and mode mismatching as well.

How is the density matrix of the light mode affected by the absorber? Let us translate our simple beam-splitter picture into mathematical terms. The reduced density operator $\hat{\rho}(\eta)$ of the split signal $\hat{\rho}$ is given by

$$\hat{\rho}(\eta) = \text{tr}_2\{\hat{B}^\dagger(\eta)\hat{\rho}|0\rangle_2\langle 0|_2\hat{B}(\eta)\}. \tag{4.54}$$

Here tr_2 denotes the trace with respect to the second mode involved (the absorbed part of the signal). The operator $\hat{B}(\eta)$ describes a real beam-splitter transformation (4.20) with the parameters $\cos(\Theta/2) = \eta^{1/2}$ and $\Phi = \Psi = \Lambda = 0$. The second input field of the absorber (the fluctuation reservoir) is in the vacuum state $|0\rangle_2$. We calculate the trace tr_2 in Eq. (4.54) in the Fock basis and obtain the formula

$$\langle m|\hat{\rho}(\eta)|n\rangle = \sum_{k=0}^{\infty} \langle m, k|\hat{B}^\dagger(\eta)|0\rangle_2\hat{\rho}\langle 0|_2\hat{B}(\eta)|n, k\rangle \tag{4.55}$$

for the partially absorbed light in the Fock representation. The required matrix elements are easily taken from the general formula (4.42) describing the optical interference of Fock states. We obtain

$$\langle 0|_2\hat{B}(\eta)|n, k\rangle = \left[b_n^{n+k}(\eta)\right]^{1/2}|n + k\rangle \tag{4.56}$$

where $b_n^m(\eta)$ denotes the binomial distribution (also called the Bernoulli distribution)

$$b_n^m(\eta) = \binom{m}{n}\eta^n(1 - \eta)^{m-n}. \tag{4.57}$$

Consequently, the density matrix of the partially absorbed light is given by the expression

$$\langle m|\hat{\rho}(\eta)|n\rangle = \sum_{k=0}^{\infty} \left[b_n^{n+k}(\eta)b_m^{m+k}(\eta)\right]^{1/2}\langle m + k|\hat{\rho}|n + k\rangle \tag{4.58}$$

in terms of the density matrix $\langle m|\hat{\rho}|n\rangle$ of the initial field. This formula describes a *generalized Bernoulli transformation* [136], [148]. To understand the principal effect of this transformation, let us consider the diagonal elements only, that is, the photon-number distribution $p_n(\eta) = \langle n|\hat{\rho}(\eta)|n\rangle$ in terms of the initial photon statistics $p_n = \langle n|\hat{\rho}|n\rangle$. We obtain from Eq. (4.58)

$$p_n(\eta) = \sum_{m=n}^{\infty} \binom{m}{n} \eta^n (1-\eta)^{m-n} p_m. \qquad (4.59)$$

This formula shows that during absorption, photons act like classical particles, a phenomenon familiar from the photon lottery in beam splitting. See Section 4.1.3. Because energy is ultimately lost and not gained, states $|m\rangle$ with fewer photons than n can by no means produce n photons. Consequently, the series (4.59) begins at $m = n$. In order to transmit n photons from $|m\rangle$ the absorber chooses any n photons at random with probability η for each. Simultaneously, precisely $m - n$ photons are selected for absorption, each with probability $(1 - \eta)$. Because the photon statistics $p_n(\eta)$ does not discriminate which particular photon has contributed to it, the transition probability is enhanced by the number of all possible choices, that is, by the binomial coefficient $\binom{m}{n}$. In this way all photon numbers with $m \geq n$ (occurring with probabilities p_m) influence the photon statistics of the partially absorbed field.

In another way of understanding the effect of the absorber on the photon-number distribution, let us assume that the light is gradually absorbed. The coefficient η depends exponentially on a parameter l that might be an appropriately scaled penetration depth, for instance. We use the parameterization (Beer's law)

$$\eta = \exp(-l) \qquad (4.60)$$

and obtain from the result (4.59) the master equation

$$\frac{dp_n(l)}{dl} = -np_n(l) + (n+1)p_{n+1}(l). \qquad (4.61)$$

On average a fraction of $(n + 1)dl$ photons jump from $|n + 1\rangle$ to $|n\rangle$ and ndl photons leave the state $|n\rangle$ within dl. The absorption of light is very similar to the spontaneous emission of light from an atomic oscillator. Here, however, the role of matter and light is reversed – the light mode "emits spontaneously" and the material absorber swallows the "emitted" energy quanta.

4.2 Homodyne detector

Photodetection is the very basis of quantum-optical measurements. Real experiments with nonclassical light require highly efficient detectors at the limits of present-day technology. These detectors must have low electronic noise, and they should have single-quantum resolution. So any progress in detector design widens the scope of measuring interesting quantum effects. How do detectors work, and which types are commonly used?

4.2.1 Photodetector

Most photodetectors apply a version of the *photoelectric effect* to operate (which was discovered by Hertz [114] in 1887 and has been important for quantum mechanics since Einstein's 1905 paper [83]). Radiation ionizes a piece of photosensitive material and produces freely moving electrons, that is, an electric current that can be amplified and handled by electronic means. In *avalanche photodiodes* operating in the Geiger mode, electrons of the valence band are lifted into the conduction band by absorption of radiation. (Of course, holes are created as well, which we do not consider here for simplicity.) As in the famous Geiger–Müller detector, a voltage (about 200 volts) is applied to accelerate the conducting electrons. By collisions, they bring other charge carriers into the conduction band so that an avalanche is formed and a distinct "click" is produced. Under ideal circumstances this click is caused by the absorption of a single photon lifting one electron into the conduction band, which is then multiplied. It is evident from this simple picture that the click might be also a "false alarm." An electron being by chance in the conduction band will trigger an avalanche as well. Despite this possibility, present state-of-the-art avalanche photodiodes may reach about eighty percent quantum efficiency [142] without significant false counts.

However, another problem is involved in avalanche photodetection. The current is not proportional to the photon number because the growth of the carrier avalanche is not limited by the number of detected photons but by the saturation of the amplification process. Avalanche photodiodes do not or do click, but they cannot display how many photons are contained in a single click. However, a way exists to measure the photon statistics if the field is sufficiently weak and constant during a certain time window. Photocounting is performed by counting electronically the number of detector clicks in the given time interval. This method (*time-correlated single-photon counting* [72]) works because the signal is weak enough so that the probability of detecting simultaneously two photons is very low. By repeating this procedure many times, the photon statistics of the investigated light is measured. Alternative photodetectors that can discriminate between single photons are photomultipliers [88] and streak cameras [73]. However, the efficiency of these detectors reaches only ten to twenty percent and the single-photon resolution is lost when more than about ten photons are detected.

Another commonly used detector type is the linear-response photodiode. In most cases, the photosensitive part of the detector is a *p-i-n* structure, a sandwich of p(ositively) doped, i(ntrinsic), and n(egatively) doped semiconductor material. Commonly, Si or InGaAs are used where Si detects light out to a one-micrometer wavelength and InGaAs operates in the range 1.0 to 1.1 micrometers. A bias voltage (about ten volts) is applied to drain the majority carriers (electrons in *n* and holes in *p*) out of the intrinsic zone. In this depletion region an unstable situation is created for the minority carriers. As soon as electron-hole pairs are present in the intrinsic zone, the bias voltage produces a current that is proportional to the number of carriers. As in the previously discussed avalanche photodiode, light creates electron-hole pairs in the depletion zone. This process can be made highly efficient [221] because in contrast to the avalanche photodiode, the applied voltage is low, so that no avalanche is forming. The current response of the detector is linear in the intensity of the detected light. On the other hand, thermal fluctuations cause Nyquist noise in the photocurrent. In addition, thermal effects create naturally electron-hole pairs in the depletion zone, producing the so-called *dark current*. Because of this electronic noise, linear-response photodiodes do not reach single-photon resolution. They are suitable for relatively high intensities (greater than about 100 photons per microsecond).

In any case, we must face inefficiencies and noise in realistic photodetection. A convenient model to understand the effect of these experimental problems is provided by imagining a fictitious beam splitter placed in front of an ideal detector. See Fig. 4.4. Only the transmitted photons are counted, so that the transmissivity of the beam splitter corresponds to the detection efficiency. Dissipation is always accompanied by fluctuations. These degrade the quantum-noise properties of the detected light. The fluctuations are modeled by a vacuum entering the unused port of the fictitious beam splitter. We have analyzed in Section 4.1.4 how the Wigner function of the signal is smoothed during absorption. This analysis shows how the nonclassical features of light are lost when the detectors are inefficient.

4.2.2 Balanced homodyne detection

Under idealized circumstances the photon number is measured in direct photodetection. However, another method of detection exists, in which the field amplitudes (the quadrature components) are measured instead of the quantized intensity. As we have seen in Chapter 2, intensity (photon number) and field amplitude (quadrature) are distinct quantities. There is no simple relationship between the photon statistics and the quadrature distributions in the quantum regime. (We will show in Section 5.2 how the two are related to each other.) Additionally, the field amplitudes contain phase information, and so they are dependent on phase. Quadrature components $\hat{q}_\theta$ are defined (2.15) with respect to a certain reference phase θ that can be varied experimentally. How do we measure the quadratures?

In 1983 Yuen and Chan [302] proposed and subsequently Abbas, Chan, and Yee [1] first demonstrated balanced homodyne detection, a method designed for measuring the degree of quadrature squeezing. (Note that there are also other ways of quadrature measurements based on earlier versions of homodyne detection [252], [304], [305]. The balanced version of homodyne detection has the great practical advantage of canceling technical noise and the classical instabilities of the reference field. Note also that the idea of homodyne detection was born in the microwave technology developed during World War II. The precursor of the optical balanced homodyne detector was the microwave balanced-mixer radiometer [75].)

The principal scheme of the balanced homodyne detector is depicted in Fig. 4.5. The signal interferes with a coherent laser beam at a well-balanced 50:50 beam splitter. The laser-light field is called the *local oscillator* (LO). It provides the phase reference θ for the quadrature measurement. (We assume that the signal and the LO have a fixed phase relation, as is the case in most experiments applying homodyne detection, because both fields are ultimately generated by a common master laser.) The local oscillator should be intense

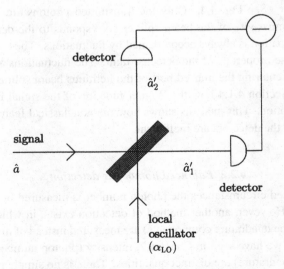

Fig. 4.5. Diagram of a balanced homodyne detector.

with respect to the signal for providing a precise phase reference. We will assume that the LO is powerful enough to be treated classically, that is, we neglect totally the quantum fluctuations of the LO. After the optical mixing of the signal with the local oscillator, each emerging beam is directed to a photodetector (usually a linear-response photodiode). The photocurrents I_1 and I_2 are measured, electronically processed, and finally subtracted from each other. The difference $I_{21} \equiv I_2 - I_1$ is the quantity of interest because it contains the interference term of the LO and the signal. We assume for simplicity that the measured photocurrents I_1 and I_2 are proportional to the photon numbers $\hat{n}_1$ and $\hat{n}_2$ of the beams striking each detector. They are given by

$$\hat{n}_1 = \hat{a}_1'^{\dagger} \hat{a}_1' \quad \text{and} \quad \hat{n}_2 = \hat{a}_2'^{\dagger} \hat{a}_2' \tag{4.62}$$

in terms of the mode operators

$$\hat{a}_1' = 2^{-1/2}(\hat{a} - \alpha_{Lo}), \quad \hat{a}_2' = 2^{-1/2}(\hat{a} + \alpha_{Lo}) \tag{4.63}$$

of the fields emerging from the beam splitter. Here $\hat{a}$ denotes the annihilation operator of the signal, and α_{Lc} the complex amplitude of the local oscillator. (As already mentioned, we treat the LO classically.) The difference I_{21} is proportional to the difference photon number (assuming perfect quantum efficiency)

$$\hat{n}_{21} = \hat{n}_2 - \hat{n}_1 = \alpha_{Lo}^* \hat{a} + \alpha_{Lo} \hat{a}^{\dagger}. \tag{4.64}$$

We denote the phase of the local oscillator by θ and see immediately from the definition (2.15) of the quadrature components that the measured quantity I_{21} is indeed proportional to $\hat{q}_\theta$ because

$$\hat{n}_{21} = 2^{1/2} |\alpha_{Lo}| \hat{q}_\theta. \tag{4.65}$$

A balanced homodyne detector measures the quadrature component $\hat{q}_\theta$. The reference phase θ is provided by the local oscillator and can be varied by adjusting the LO using, for instance, a piezo-electrically movable mirror. An experimental method for finding the scaling of the quadrature component in the difference current I_{21} is to keep a record of the sum current because the sum of I_2 and I_1 is proportional to $|\alpha_{Lo}|^2$ in leading order. This can be experimentally important because the intensity of the local oscillator is usually an unknown quantity.

Although our analysis is rather crude, the final result (4.65) is indeed correct and has been verified by more sophisticated theories of homodyne detection [39], [54], [208], [229], [280], [281]. The reason is that in the regime of a strong local oscillator, classical optics is well-justified for understanding the behavior of the LO field (in the sense of Bohr's correspondence principle). Our simple model is typical for any other theories of quantum measurement. A

quantum object, here the signal mode, encounters a classical apparatus, here
the beam splitter, the local oscillator, and the detectors. The theoretical model
of the measurement is a hybrid of quantum mechanics and classical physics.
We may certainly push the limits between the quantum world and the classical
apparatus a bit further. We may quantize the local oscillator, for instance, but
there is always a border between measuring device and quantum system. It is
wise to make the cut between the quantum world and the classical apparatus as
early as possible yet without losing essential features.

The balanced homodyne detector shares another feature of many other
quantum-measurement devices – the detector is an amplifier. The local oscillator amplifies the signal by the mutual optical mixing of the two. Or, seen
from a different point of view, the homodyne detector is an interferometer that
can be measurably imbalanced by a single photon in the signal mode, because
the reference field is very intense. The amplification has an important technical
advantage. The so-amplified signal is well above the electronic noise floor
of the photodiodes. The signal amplitude is enhanced so that even the noisy
linear-response photodiodes can detect the quantum features of the signal. On
the other hand, these photodetectors may reach nearly 100 percent efficiency
[221]. In this way the balanced homodyne detector takes advantage of the
high efficiency of photodiodes and at the same time can determine signals with
single-photon resolution – a nearly perfect technical solution!

4.2.3 Spatial–temporal modes

Because the local oscillator serves as a coherent amplifier, it also chooses the
signal mode. The LO selects only the spatial–temporal mode of the continuous
quantum field "light" that matches the local-oscillator field. In this way the
observer separates the quantum object (a single optical mode) from the "rest of
the world." Studying this issue a bit more carefully requires a brief excursion
into the quantum-field theory of light. This theory, quantum electrodynamics
(QED), is one of the most beautiful and successful theoretical models in physics.
QED has become the paradigm for any relativistic gauge field theory. Many
excellent books cover the subject; see for instance Refs. [28], [32], [59], [95],
[108], [109], [122]. However, for our purpose we do not need quantum electrodynamics in its full glory. Instead, we sketch the quantum theory of a typical
optical approximation.

In classical optics the wave field is often represented by a complex-valued
and scalar function. The polarization is fixed, magnetic effects are neglected,
and the (complex) positive-frequency part $E^{(+)}$ of the electric field strength is
considered the field quantity. Note that the negative frequency part $E^{(-)}$ is just

the complex conjugate of $E^{(+)}$. The electric energy density $E^2/2 = (E^{(+)} + E^{(-)})^2/2$ contains two rapidly oscillating terms $E^{(+)2}$ and $E^{(-)2}$ for fields in the optical frequency range (hundreds of terahertz). Because photodetection requires a temporal integration, these oscillations cannot be observed. As a consequence the quantity $|E^{(+)}|^2$ rather than the electric energy density $E^2/2$ is regarded as the intensity of the field. In this approximation we leave certain features (and problems) out of consideration in order to focus on the most relevant aspects in the optical domain of the spectrum.

In the quantum theory of scalar optics we represent the field by the operator $\hat{E}^{(+)}(x, t)$ for the positive-frequency component of the electric field strength at a certain polarization. The three-dimensional vector $x = (x, y, z)$ describes a spatial point, whereas t denotes the time. The quantum field "light" comprises the wealth of possibilities allowed by the field equation, which is, in our case, the wave equation in free space

$$\left(\frac{\partial^2}{\partial x^2} + \frac{\partial^2}{\partial y^2} + \frac{\partial^2}{\partial z^2} - \frac{1}{c^2}\frac{\partial^2}{\partial t^2}\right)\hat{E}^{(+)}(x, t) = 0. \qquad (4.66)$$

Here c denotes the speed of light. Because this equation is linear, we can always represent the field operator $\hat{E}^{(+)}(x, t)$ in terms of a mode expansion

$$\hat{E}^{(+)}(x, t) = \sum_k \omega_k^{1/2}\hat{a}_k v_k(x)\exp(-i\omega_k t). \qquad (4.67)$$

We require that the spatial mode functions $v_k(x)$ are subject to the Helmholtz equation

$$\left(\frac{\partial^2}{\partial x^2} + \frac{\partial^2}{\partial y^2} + \frac{\partial^2}{\partial z^2} + \frac{\omega_k^2}{c^2}\right)v_k(x) = 0 \qquad (4.68)$$

with the positive frequency ω_k. In this way we guarantee that $\hat{E}^{(+)}(x, t)$ satisfies the wave equation (4.66). Because the Laplace operator $\partial^2/\partial x^2 + \partial^2/\partial y^2 + \partial^2/\partial z^2$ is Hermitian, the eigenfunctions (4.68) are orthonormal

$$\int_{-\infty}^{+\infty}\int_{-\infty}^{+\infty}\int_{-\infty}^{+\infty} v_k^*(x)v_{k'}(x)\,dx\,dy\,dz = \delta_{kk'}. \qquad (4.69)$$

Moreover, we can find a complete set of spatial mode functions

$$\sum_k v_k^*(x)v_k(x') = \delta(x - x')\delta(y - y')\delta(z - z'). \qquad (4.70)$$

The completeness relation means that the field operator accounts for all possible optical superpositions of field excitations. Given a certain set of mode functions, the only free quantities of the field are the "expansion coefficients" $\hat{a}_k$. We postulate that the energy density of the fields is the local intensity

$\hat{E}^{(+)}(x, t)^{\dagger} \hat{E}^{(+)}(x, t)$ and obtain from the orthonormality relation (4.69) that the total energy

$$\hat{H} = \int_{-\infty}^{+\infty} \int_{-\infty}^{+\infty} \int_{-\infty}^{+\infty} \hat{E}^{(+)}(x, t)^{\dagger} \hat{E}^{(+)}(x, t) \, dx \, dy \, dz \qquad (4.71)$$

is given as the sum

$$\hat{H} = \sum_{k} \omega_k \hat{a}_k^{\dagger} \hat{a}_k. \qquad (4.72)$$

We interpret the energy $\hat{H}$ as the Hamiltonian of the field. In other words, we require the dynamics of the light field to be governed by the Heisenberg equation

$$i\frac{\partial}{\partial t} \hat{E}^{(+)}(x, t) = \left[\hat{E}^{(+)}(x, t), \hat{H} \right]. \qquad (4.73)$$

As already mentioned, the mode expansion (4.67) together with the Helmholtz equation (4.68) guarantees already that the field operator $\hat{E}^{(+)}(x, t)$ obeys the wave equation (4.66). However, we seek certain requirements on the "expansion coefficients" $\hat{a}_k$ such that the dynamical law for $\hat{E}^{(+)}(x, t)$ is the Heisenberg equation (4.73). For this we substitute the mode expansions (4.67) and (4.72) in Eq. (4.73) to obtain for each component

$$\omega_{k'} \hat{a}_{k'} = \left[\hat{a}_{k'}, \sum_{k} \omega_k \hat{a}_k^{\dagger} \hat{a}_k \right] \qquad (4.74)$$

because the spatial mode functions are linearly independent. In order to satisfy (4.74) we require that the "expansion coefficients" be independent bosonic annihilation operators obeying the commutation relation

$$\left[\hat{a}_{k'}, \hat{a}_k^{\dagger} \right] = \delta_{k'k}. \qquad (4.75)$$

Photons are bosons. This constraint is the only one imposed on the light field obeying the wave equation (4.66).

We have sketched the key elements of quantum optics as a field theory. The optical field consists of infinitely many harmonic oscillators. The total energy is the sum (4.72) of the mode energies apart from the *vacuum energy* $\sum_{k} \omega_k / 2$, which is infinite for infinitely many modes. This problem is one of those we have circumvented in our simplified approach.

Photodiodes measure the flux of photons rather than the energy density of the field. This measurement is conveniently described by introducing the *flux operator*

$$\hat{\phi}(x, t) = \sum_{k} \hat{a}_k v_k(x) \exp(-i\omega_k t). \qquad (4.76)$$

We obtain from the completeness (4.70) of the mode functions $v_k(x)$ and from the bosonic commutation relation (4.75) that the flux operators are bosonic as well

$$[\hat{\phi}(x, t), \hat{\phi}^\dagger(x', t)] = \delta(x - x')\delta(y - y')\delta(z - z'). \quad (4.77)$$

The flux $\hat{\phi}(x, t)$ plays the role of a *local* annihilation operator at the spatial point x. The local flux density $\hat{\phi}^\dagger(x, t)\hat{\phi}(x, t)$ accounts for the photons localized in a detection process. (Note, however, that this way of *localizing photons* relies on our simplified scalar model for the optical field. We have entirely neglected the transversal nature of light, which causes serious problems in the concept of photon localization. See for instance Ref. [187], Section 12.11. This difficulty is yet another one we have circumvented in our approach.) The total flux

$$\int_{-\infty}^{+\infty}\int_{-\infty}^{+\infty}\int_{-\infty}^{+\infty} \hat{\phi}^\dagger(x, t)\hat{\phi}(x, t)\,dx\,dy\,dz = \sum_k \hat{a}_k^\dagger \hat{a}_k \quad (4.78)$$

gives the total number of photons. The description of localized photons by the flux density $\hat{\phi}^\dagger(x, t)\hat{\phi}(x, t)$ motivates us to assume that the measured photocurrent is proportional to the flux $\hat{n}$ at the detector surface D integrated during the time interval $[0, T]$, with

$$\hat{n} = \int_0^T \iint_D \hat{\phi}^\dagger(x_D, t)\hat{\phi}(x_D, t)\,dx_D\,dy_D\,dt. \quad (4.79)$$

Here we are mainly interested in *beams* traveling toward the detector. So we assume that the field propagates chiefly along one direction, say, along the z axis. What does this assumption mean? Any local field quantity $\hat{F}$ at the position z and the time $t + t'$ has just propagated from the position $z - ct'$ at the time t, that is,

$$\hat{F}(z, t + t') = \hat{F}(z - ct', t). \quad (4.80)$$

Strictly speaking, the field equation (4.66) does not allow this way of straight propagation simply because we have broken the spatial symmetry of (4.66) by distinguishing the z axis. The propagation relation (4.80) cannot be universally valid. It is restricted to those states of the quantum field "light" that realize a beamlike propagation along the z axis. In classical optics this behavior is one feature of the *paraxial approximation* [258]. Using the propagation (4.80) we obtain the commutation relation

$$[\hat{\phi}(x, t), \hat{\phi}^\dagger(x', t')] = \delta(x - x')\delta(y - y')\delta(ct - ct')$$

$$= \delta(x - x')\delta(y - y')\frac{1}{c}\delta(t - t') \quad (4.81)$$

for the flux operators at identical spatial planes $z = z'$ (at the detector surface $z = z_D$, for instance) but at different times.

Let us now turn to the description of balanced homodyne detection in our simple quantum field theory of light. The signal interferes with the local oscillator at a well-balanced 50 : 50 beam splitter, and the emerging beams travel to two photodiodes. The fields $\hat{E}_1^{(+)}$ and $\hat{E}_2^{(+)}$ at the detectors are given by

$$\hat{E}_1^{(+)} = 2^{-1/2}\big(\hat{E}_s^{(+)} - \hat{E}_{LO}^{(+)}\big),$$
$$\hat{E}_2^{(+)} = 2^{-1/2}\big(\hat{E}_s^{(+)} + \hat{E}_{LO}^{(+)}\big). \tag{4.82}$$

In this formula we have distinguished the signal $\hat{E}_s^{(+)}$ and the local oscillator $\hat{E}_{LO}^{(+)}$ as two parts of the quantum field. Possible phase shifts brought about by the beam splitter are incorporated in the reference phases used to define the signal and the local oscillator. Furthermore, we have assumed that the beam splitter is balanced at the relevant frequency range and that the propagation after the optical mixing is identical for the emerging beams hitting the detector surfaces. Because the beam splitter is frequency-independent, the flux operators $\hat{\phi}_s$ and $\hat{\phi}_{LO}$ for the signal and the local oscillator are mixed in the same way as the field–strength components $\hat{E}_s^{(+)}$ and $\hat{E}_{LO}^{(+)}$, that is,

$$\hat{\phi}_1 = 2^{-1/2}(\hat{\phi}_s - \hat{\phi}_{LO}), \quad \hat{\phi}_2 = 2^{-1/2}(\hat{\phi}_s + \hat{\phi}_{LO}). \tag{4.83}$$

In balanced homodyne detection the quantity of interest is the difference of the measured photocurrents. The local oscillator should be intense enough so that the currents are well above the electronic noise floor of the diodes. In this case the photocurrent is proportional to the photon number given by Eq. (4.79). We use the description (4.83) of the optical interference at the beam splitter and obtain for the photon-number difference

$$\hat{n}_{21} = \int_0^T \iint_D [\hat{\phi}_{LO}^\dagger \hat{\phi}_s + \hat{\phi}_s^\dagger \hat{\phi}_{LO}] \, dx_D \, dy_D \, dt. \tag{4.84}$$

The local oscillator is intense and coherent, so that we can replace the flux $\hat{\phi}_{LO}$ by a classical field

$$\hat{\phi}_{LO} \approx \phi_{LO}(x, t) = \alpha_{LO} u(x, t). \tag{4.85}$$

We choose α_{LO} so that the function $u(x, t)$ is normalized according to

$$\int_0^T \iint_D |u(x_D, t)|^2 \, dx_D \, dy_D c \, dt = 1. \tag{4.86}$$

In this way $|\alpha_{LO}|^2$ quantifies the total "mean photon number" of the local oscillator in the light volume that has traveled to the detector during the time T, whereas $u(x, t)$ describes the spatial–temporal beam shape. We define the operator

$$\hat{a} \equiv \int_0^T \iint_D \hat{\phi}_s(x_D, t) u^*(x_D, t) \, dx_D \, dy_D \, dt \tag{4.87}$$

and arrive at our familiar formula (4.64) for the photon-number difference $\hat{n}_{21}$. The notation suggests already that $\hat{a}$ is an annihilation operator. Yet this property is nontrivial, and it is the final fruit of our excursion into the quantum field theory of light. We use the commutation relation (4.81) of the flux operators for beamlike propagation and the normalization (4.86) of $u(x, t)$ to get the result

$$[\hat{a}, \hat{a}^\dagger] = 1. \qquad (4.88)$$

This relation is the very origin for the simple quantum theory of light developed in Chapter 2. The whole machinery of single-mode quantum optics follows from the commutation relation (4.88). Note, however, that the formula (4.88) relies critically on the relation (4.81). Strictly speaking, our result is only an approximation and is valid only for beamlike propagation.

We have seen that the local oscillator singles out one bosonic mode from the rest of the quantum field "light." The mode function is given by the spatial–temporal shape of the local-oscillator beam at the detector surface and during the time interval $[0, T]$. The overall phase and intensity of the LO is comprised in the complex amplitude α_{LO}. Shifting the phase $\theta = \arg \alpha_{LO}$ rotates the measured quadratures $\hat{q}_\theta$. The observer defines via the local oscillator the frame in space and time that is subject to the field–quadrature measurement. By tailoring the shape of the LO beam [112], [285], [286] high spatial–temporal resolution can be achieved.

On the other hand, what happens if the signal we wish to observe is not completely localized in the selected mode? Suppose that the flux $\hat{\phi}_s$ consists of two parts: one excited spatial–temporal mode $\hat{a}_s u_s(x, t)$ and a vacuum field $\hat{\phi}_0$ that accounts for all other potential modes in $\hat{\phi}_s$, that is,

$$\hat{\phi}_s = \hat{a}_s u_s(x, t) + \hat{\phi}_0. \qquad (4.89)$$

The mode $\hat{a}_s u_s(x, t)$ with the bosonic annihilation operator $\hat{a}_s$ is the signal we wish to observe. We define the quantity

$$\mu_M^{1/2} \equiv \int_0^T \iint_D u_s(x_D, t) u^*(x_D, t) \, dx_D \, dy_D \, dt. \qquad (4.90)$$

Note that we can always adjust the overall phase of u_s in such a way that η_M is a nonnegative real number. We use the Schwarz inequality and the normalization (4.86) of u and u_s to obtain the bound

$$\eta_M \leq 1. \qquad (4.91)$$

Furthermore, we define the operator $\hat{a}_2$ according to the relation

$$(1 - \eta_M)^{1/2} \hat{a}_2 = \int_0^T \iint_D \hat{\phi}_0(x_D, t) u^*(x_D, t) \, dx_D \, dy_D \, dt. \qquad (4.92)$$

Consequently, the annihilation operator $\hat{a}$ of the spatial–temporal mode that matches the local-oscillator beam is given by the familiar-looking expression

$$\hat{a} = \eta_M^{1/2}\hat{a}_s + (1 - \eta_M)^{1/2}\hat{a}_2. \tag{4.93}$$

We obtain from the bosonic commutation relations of $\hat{a}_s$ and $\hat{a}$ that $\hat{a}_2$ must obey $[\hat{a}_2, \hat{a}_2^\dagger] = 1$, that is, that $\hat{a}_2$ is a bosonic annihilation operator as well. Because $\hat{\phi}_0$ describes a vacuum field, the operator $\hat{a}_2$ corresponds to a vacuum mode. So we see that the effect of *mode mismatching* is easily described in terms of a simple beam-splitter model. The transmitted beam is observed while the "reflected" light is lost. The efficiency η_M defined in Eq. (4.90) quantifies the degree of mode matching between signal and local oscillator. In particular, η_M reaches unity if the mode overlap (4.90) is perfect.

4.2.4 Inefficiencies in homodyne detection

Photodetection is usually not completely efficient in practice. In order to understand the influence of inefficiencies on homodyne detection, we use our simple model for losses in direct photodetection. We imagine fictitious beam splitters to be placed in front of the two detectors (assumed ideal) in the measurement setup (see Fig. 4.6). For the annihilation operators of the detected fields, we find according to Eq. (4.53)

$$\hat{a}_1'' = \eta^{1/2}\hat{a}_1' + (1 - \eta)^{1/2}\hat{b}_1, \quad \hat{a}_2'' = \eta^{1/2}\hat{a}_2' + (1 - \eta)^{1/2}\hat{b}_2. \tag{4.94}$$

Here $\hat{b}_1$ and $\hat{b}_2$ denote the annihilation operators of the vacuums entering the second ports of the fictitious beam splitters. We obtain for the photon-number difference

$$
\begin{aligned}
\hat{n}_{21} &= \hat{a}_2''^\dagger\hat{a}_2'' - \hat{a}_1''^\dagger\hat{a}_1'' \\
&= \left(\eta^{1/2}\hat{a}_2'^\dagger + (1 - \eta)^{1/2}\hat{b}_2^\dagger\right)\left(\eta^{1/2}\hat{a}_2' + (1 - \eta)^{1/2}\hat{b}_2\right) \\
&\quad - \left(\eta^{1/2}\hat{a}_1'^\dagger + (1 - \eta)^{1/2}\hat{b}_1^\dagger\right)\left(\eta^{1/2}\hat{a}_1' + (1 - \eta)^{1/2}\hat{b}_1\right) \\
&= \eta\left(\hat{a}_2'^\dagger\hat{a}_2' - \hat{a}_1'^\dagger\hat{a}_1'\right) \\
&\quad + [\eta(1 - \eta)]^{1/2}\left(\hat{a}_2'^\dagger\hat{b}_2 + \hat{b}_2^\dagger\hat{a}_2' - \hat{a}_1'^\dagger\hat{b}_1 - \hat{b}_1^\dagger\hat{a}_1'\right) \\
&\quad + (1 - \eta)\left(\hat{b}_2^\dagger\hat{b}_2 - \hat{b}_1^\dagger\hat{b}_1\right).
\end{aligned}
\tag{4.95}
$$

The annihilation operators $\hat{a}_1'$ and $\hat{a}_2'$ given by Eq. (4.63) describe the fields emerging from the 50 : 50 beam splitter where the signal is optically mixed with the local oscillator. The LO should be a rather intense coherent field compared with the signal. We treat the local oscillator classically and consider

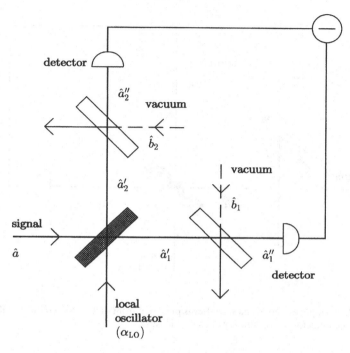

Fig. 4.6. Fictitious beam splitters are placed in front of the (assumed ideal) photodetectors in a balanced homodyne detection scheme to account for detection losses.

only the leading terms in Eq. (4.95) with respect to α_{LO}. In this way we obtain the simple result

$$\hat{n}_{21} = \eta^{1/2}\alpha_{LO}^*(\eta^{1/2}\hat{a} + (1-\eta)^{1/2}\hat{b}) + \text{H. c.} \qquad (4.96)$$

(The symbol H. c. denotes the Hermitian conjugate of the other part of an expression.) We have defined the annihilation operator

$$\hat{b} \equiv 2^{-1/2}(\hat{b}_2 - \hat{b}_1). \qquad (4.97)$$

This operator corresponds to the optical mixing of the fictitious vacuum-noise modes $\hat{b}_1$ and $\hat{b}_2$, and it obeys the bosonic commutation relation $[\hat{b}, \hat{b}^\dagger] = 1$. Because the interference of vacuum with vacuum yields vacuum, the fluctuation mode $\hat{b}$ can be regarded as a bosonic mode, being in the vacuum state as well.

Formula (4.96) has a simple interpretation: Similar to direct photon counting, a fictitious vacuum field has to be added to the intensity-reduced signal in homodyne detection. This interpretation means that we may replace the arrangement of two fictitious beam splitters in front of the photodetectors (see

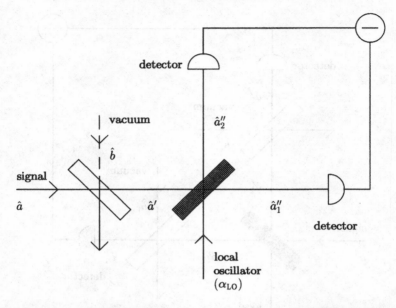

Fig. 4.7. A single effective beam splitter is placed in front of a (assumed ideal) balanced homodyne detector to account for detection losses.

Fig. 4.6) by just one *effective* beam splitter in front of an ideal homodyne detector (see Fig. 4.7). This effective beam splitter accounts for other kind of losses as well, in particular for mode mismatch. Consequently, we can use the theory developed in Section 4.1 to predict the quantum effects of both detection losses and mode mismatch comprised in an effective η.

The measured distribution is the quadrature distribution for a mode that has been attenuated to some extent. To be more quantitative, we consider the absorption in the Wigner representation and use our basic axiom (3.1) that the marginals of the Wigner function yield the quadrature distributions. All necessary mathematical ingredients have been given already. We recall that the Wigner function is transformed into an s-parameterized quasiprobability distribution (4.48) with s given by Eq. (4.49). According to Eq. (3.67) the marginal distributions, that is, the measured quadrature histograms $\mathrm{pr}(q, \theta; \eta)$, are given in terms of the ideal quadrature distributions $\mathrm{pr}(q, \theta)$ by the formula

$$\mathrm{pr}(q, \theta; \eta) = (\pi(1 - \eta))^{-1/2} \int_{-\infty}^{+\infty} \mathrm{pr}(x, \theta) \exp\left[-\frac{\eta}{1 - \eta}(x - \eta^{-1/2}q)^2\right] dx.$$

$$(4.98)$$

A realistic homodyne detector measures an appropriately smoothed quadrature distribution. We remark that more refined quantum-statistical theories of homodyne detection [229], [280] yield the same result.

4.3 Further reading

The first quantum theory describing the action of a loss less beam splitter was developed by W.H. Brunner, H. Paul, and G. Richter [44], [209] starting from a microscopic model; for a review see Ref. [211]. See also the papers by Y. Aharonov, D. Falkoff, E. Lerner, and H. Pendleton [5]; A. Zeilinger [306]; S. Prasad, M.O. Scully, and W. Martienssen [223]; Z.Y. Ou, C.K. Hong, and L. Mandel [206]; B. Huttner and Y. Ben-Aryeh [121]; J. Janszky, P. Adam, M. Bertolotti, C. Sibilia, and Y. Yushin [127], [128], [129]; V. Peřinová, A. Lukš, J. Křepelka, C. Sibilia, and M. Bertolotti [218]; W.K. Lai, V. Bužek, and P.L. Knight [143]; L. Allen and S. Stenholm [7]; R.W.F. van der Plank and L.G. Suttorp [276]; and A. Luis and L.L. Sánchez Soto [182]. Probably the most detailed papers on the quantum theory of beam splitting are the article [53] by R.A. Campos, B.E.A. Saleh, and M.C. Teich and Ref. [155]. Extensions to multiports are considered by M. Reck, A. Zeilinger, H.J. Bernstein, and P. Bertani [230] and P. Törmä, I. Jex, S. Stenholm, and A. Zeilinger [131], [260], [268], [269]. The quantum theory of other (theoretically) simple optical instruments (linear amplifiers, nondegenerate parametric amplifiers, and phase-conjugating mirrors) is studied in Refs. [151], [157], and in the references cited therein.

Recent theories of homodyne detection were developed by S.L. Braunstein [39]; W. Vogel and J. Grabow [280]; Z.Y. Ou and H.J. Kimble [208]; and by M.G. Raymer, J. Cooper, H.J. Carmichael, M. Beck, and D.T. Smithey [229]. The latter paper is probably the most comprehensive. Here the concept of spatial–temporal modes is developed, and important practical issues of balanced-homodyne detection are discussed. The effect of detection inefficiencies was considered in detail in Refs. [156], [159], [160]. It is also interesting to look over "early" papers on homodyne detection because they contain a number of later rediscovered results; see for instance the classic trilogy [252], [304], [305] by H.P. Yuen, J.H. Shapiro, and J.A. Machado Mata and the papers cited in Ref. [229].

5

Quantum tomography

5.1 Phase-space tomography

As a fundamental feature of quantum mechanics, we cannot see physical objects in their full complexity. For instance, if we observe the position we cannot gain momentum information at the same time, because of Heisenberg's uncertainty principle. Quantum states may comprise complementary features that cannot be measured simultaneously and precisely. Consequently, we cannot see quantum states directly, and so the true nature of an individual quantum system is hidden for fundamental reasons. However, no principal obstacle exists to observing all complementary aspects in a series of distinct experiments on identically prepared quantum objects. Can we infer the quantum state from such a set of complementary observations? Remember that in some cases in medical imaging we cannot see an "object" directly without harming it or we are simply restricted for practical reasons, and yet we can tomographically reconstruct the shape of the hidden object. In transmission computerized tomography, for instance, a cross-section of the human body is scanned by a thin X-ray beam whose attenuated intensity is recorded by a detector. The object (or the apparatus) is rotated to yield many intensity distributions at different angles. Finally, a computer processes the data to build a picture of the object in the form of a spatial distribution of the absorption coefficient.

All we need to do for understanding the key idea of quantum-state tomography is to translate this procedure to the language of quantum mechanics. We picture the "shape" of a quantum object in phase space using the Wigner representation. The transmission profiles correspond to the marginal distributions

$$\mathrm{pr}(q, \theta) = \int_{-\infty}^{+\infty} W(q \cos \theta - p \sin \theta, q \sin \theta + p \cos \theta) \, dp \qquad (5.1)$$

of the Wigner function $W(q, p)$, that is, to shadows projected onto a line in quantum phase space. Certainly, because we cannot measure simultaneously

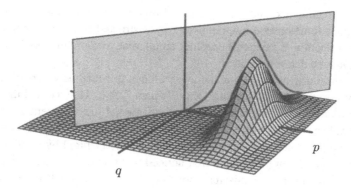

Fig. 5.1. Tomography, from the Greek word τομοσ = *slice*, is a method to infer the shape of a hidden object from its shadows (projections) under various angles. Quantum tomography is the application of this idea to quantum mechanics. For instance, in optical homodyne tomography the Wigner function plays the role of the hidden object. The observable "quantum shadows" are the quadrature distributions and are measured using homodyne detection. From these distributions the Wigner function or, more generally, the quantum state is reconstructed.

and precisely the position q and the momentum p, we cannot observe the Wigner function directly as a probability distribution. Yet we can measure the quadrature histograms

$$\mathrm{pr}(q, \theta) = \langle q | \hat{U}(\theta) \hat{\rho} \hat{U}^{\dagger}(\theta) | q \rangle, \qquad (5.2)$$

and by varying the phase θ we see the quantum object under different angles. [As usual $\hat{U}(\theta)$ denotes the phase-shifting operator defined in Eq. (2.6).] Given the distributions $\mathrm{pr}(q, \theta)$, we apply the mathematics of computerized tomography to infer the Wigner function.

In quantum optics the quadratures q_{θ} of a spatial–temporal mode can be precisely measured using balanced homodyne detection, as we have discussed in Section 4.2. Here the angle θ is defined by the phase of the local oscillator with respect to the signal. The phase θ can be easily varied using a piezo-electric translator. To measure the quadrature distributions, we may fix the phase angle θ and perform a series of homodyne measurements at this particular phase to build up a quadrature histogram $\mathrm{pr}(q, \theta)$. Then the LO phase should be changed in order to repeat the procedure at a new phase, and so on [255]. Another possibility is to monitor the phase while it drifts or to sweep it in a known way [41]. In any case, the homodyne measurement must be repeated many times on identically prepared light modes (or on a continuous-wave field) to gain sufficient statistical information about the quadrature values at a certain number of reference phases. (In Section 5.3 we develop estimations of the statistical confidence and of the required number of phase values.) Finally, the Wigner function is tomographically reconstructed from the experimental data. (In Section 5.2 we will consider an alternative sampling version of the quantum-state determination.)

The pioneering experimental demonstration of this method was published in 1993 by Smithey, Beck, Raymer, and Faridani [255]. They coined the name *optical homodyne tomography* for this fundamental measurement technique. Figure 5.2 shows the first Wigner functions reconstructed from experimental data using this scheme. The work was (indirectly) inspired by a 1989 paper of Vogel and Risken [278], who first noticed that homodyne detection can be applied for reconstructing quasiprobability distributions. Probably the first who introduced tomography to quantum mechanics were Bertrand and Bertrand [29]

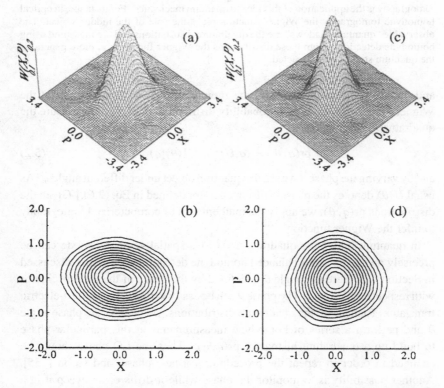

Fig. 5.2. The first "quantum pictures." Reconstructed Wigner functions of a squeezed vacuum (a, b) and a vacuum (c, d) viewed in 3D and as contour plots, with equal numbers of constant-height contours. The figures are reproduced from the pioneering paper [255] by Smithey, Beck, Raymer, and Faridani. These are the first pictures of the vacuum fluctuations and of the anisotropic quantum noise of a squeezed vacuum.

in 1987. However, the main objective of their paper was rather to give a more convincing approach to Wigner's formula (3.17) than to propose an experimental technique. (Our definition (3.1) of the Wigner function was inspired by their idea.) The mathematics of tomography dates back to 1917, when Johann Radon published in the article [266] "Über die Bestimmung von Funktionen durch ihre Integralwerte längs gewisser Mannigfaltigkeiten" an inversion formula for the later so-called *Radon transformation* (5.1). This theory was rediscovered in the 1960s and early 1970s for medical imaging. Cormack and Hounsfield shared the Nobel prize. Like any other idea of importance, quantum tomography has certain roots in the development of science.

5.1.1 Basics of tomography

How does tomography work? The ground has been already prepared in our approach to Wigner's formula (3.17) in Section 3.1. According to relation (3.9) the Fourier-transformed quadrature distribution gives the characteristic function in polar coordinates, that is, the Fourier-transformed Wigner function. We simply need to perform the Fourier inversion in polar coordinates and obtain

$$W(q, p) = \frac{1}{(2\pi)^2} \int_{-\infty}^{+\infty} \int_0^{\pi} \tilde{W}(\xi \cos\theta, \xi \sin\theta)|\xi|$$
$$\times \exp[i\xi(q\cos\theta + p\sin\theta)] \, d\theta \, d\xi$$
$$= \frac{1}{(2\pi)^2} \int_{-\infty}^{+\infty} \int_0^{\pi} \int_{-\infty}^{+\infty} \mathrm{pr}(x, \theta)|\xi|$$
$$\times \exp[i\xi(q\cos\theta + p\sin\theta - x)] \, dx \, d\theta \, d\xi \quad (5.3)$$

using Eq. (3.9) and definition (3.5) In the last line the ξ integration does not concern the quadrature distributions $\mathrm{pr}(x, \theta)$. We simplify formula (5.3), introducing the kernel

$$K(x) = \frac{1}{2} \int_{-\infty}^{+\infty} |\xi| \exp(i\xi x) \, d\xi, \quad (5.4)$$

and obtain

$$W(q, p) = \frac{1}{2\pi^2} \int_0^{\pi} \int_{-\infty}^{+\infty} \mathrm{pr}(x, \theta) K(q\cos\theta + p\sin\theta - x) \, dx \, d\theta. \quad (5.5)$$

The kernel $K(x)$ exists only in the sense of a generalized function [101] (like Dirac's delta function, which required several painful years to become accepted in mathematics). Physicists are excused, and so we can regularize the generalized function $K(x)$ in a simple way. First we express the kernel (5.4) as

$$K(x) = \frac{1}{2}\left[\int_0^{+\infty} \exp(i\xi x)\xi \, d\xi - \int_{-\infty}^0 \exp(i\xi x)\xi \, d\xi\right]$$

$$= \frac{1}{2i}\frac{\partial}{\partial x}\left[\int_0^{+\infty} \exp(i\xi x) \, d\xi - \int_{-\infty}^0 \exp(i\xi x) \, d\xi\right]$$

$$= \frac{1}{2i}\frac{\partial}{\partial x}\left[\int_0^{\infty} \exp(i\xi x) \, d\xi - \int_0^{\infty} \exp(-i\xi x) \, d\xi\right]$$

$$= \frac{\partial}{\partial x}\text{Im}\int_0^{\infty} \exp(i\xi x) \, d\xi. \tag{5.6}$$

To regularize the integral (5.6), we add an infinitely small yet positive imaginary part $+i\varepsilon$ to x so that

$$K(x) = \frac{\partial}{\partial x}\text{Im}\int_0^{\infty} \exp[i\xi(x + i\varepsilon)] \, d\xi \tag{5.7}$$

converges and obtain the result

$$K(x) = \frac{\partial}{\partial x}\text{Re}\frac{1}{x + i\varepsilon}. \tag{5.8}$$

The infinitesimal imaginary part $+i\varepsilon$ added to x means that we should slightly move the integration contour of (5.5) below the real axis. This regularization is equivalent (101) to *Cauchy's principal value* of the integral (5.5). We abbreviate this value by the symbol $\mathcal{P}$ and write

$$K(x) = \frac{\partial}{\partial x}\frac{\mathcal{P}}{x} \equiv -\frac{\mathcal{P}}{x^2}. \tag{5.9}$$

As a typical generalized function, $K(x)$ makes sense only as a kernel in integrations with respect to well-behaved functions. (Strictly speaking, $-\mathcal{P}x^{-2}$ serves only as a convenient abbreviation for the derivative of a principal-value integral involving $\mathcal{P}x^{-1}$.) Finally, we obtain the compact formula

$$W(q, p) = -\frac{\mathcal{P}}{2\pi^2}\int_0^{\pi}\int_{-\infty}^{+\infty} \frac{\text{pr}(x, \theta) \, dx \, d\theta}{(q\cos\theta + p\sin\theta - x)^2} \tag{5.10}$$

for the *inverse Radon transformation*. This expression shows how the Wigner function $W(q, p)$ can be calculated from a mathematically given set of quadrature distributions $\text{pr}(x, \theta)$. A real-world numerical application of this formula requires, however, a certain filtering, which is considered in Section 5.1.3.

5.1.2 Random phase and Abel transformation

Before we tackle the numerical issues of computerized tomography, let us study an interesting special case [158] of the Radon transformation and the inversion (5.10) What happens if we have a phase-randomized quantum state? Examples of such states are Fock states or phase-randomized coherent states. Moreover, phase-randomized quadrature distributions are obtained in homodyne measurements if the local oscillator has no fixed phase relation to the signal (for instance, if the two fields originate from different master lasers), provided of course that the drift of the relative phase is uniform. In these cases the reconstructed Wigner functions W and the quadrature distributions are invariant with respect to phase shifts. This statement means that W depends only on the radius $r = (q^2+p^2)^{1/2}$ in phase space and that all quadrature distributions are even functions and do not depend on the phase θ. Introducing polar coordinates we obtain from the Radon transformation (5.1)

$$\mathrm{pr}(x) \equiv \mathrm{pr}(x,\theta) = \int_{-\infty}^{+\infty} W(r)\,dp = 2 \int_{x}^{\infty} W(r) \frac{dp}{dr}\,dr \qquad (5.11)$$

with the momentum $p = (r^2 - x^2)^{1/2}$. The Wigner function approaches zero when the radius tends to infinity because $W(r)$ must be normalizable. Hence we obtain via partial integration

$$\mathrm{pr}(x) = -2 \int_{x}^{\infty} W'(r)(r^2 - x^2)^{1/2}\,dr. \qquad (5.12)$$

Does the inverse transformation have a similar formula? Because the Wigner function depends only on the radius r, we can replace q by r and set p to zero in the inverse Radon transformation (5.10). To perform the θ integration, we use the known integral [225], Vol. I, Eq. 2.5.16.22

$$\frac{1}{\pi} \mathrm{Re} \int_{0}^{\pi} \frac{d\theta}{r\cos\theta - x - i\varepsilon} = -\mathrm{Re}[(x^2 - r^2)^{-1/2}]$$
$$= -(x^2 - r^2)^{-1/2}\Theta(x^2 - r^2), \qquad (5.13)$$

where Θ denotes the step function. Consequently, the Wigner function is given by

$$W(r) = \frac{1}{2\pi} \int_{-\infty}^{+\infty} \mathrm{pr}(x) \frac{\partial}{\partial x}(x^2 - r^2)^{-1/2}\Theta(x^2 - r^2)\,dx. \qquad (5.14)$$

The phase-randomized quadrature distribution $\mathrm{pr}(x)$ is an even function and vanishes at infinity because $\mathrm{pr}(x)$ is normalized. We use partial integration and obtain the final result

$$W(r) = -\frac{1}{\pi} \int_{r}^{+\infty} \mathrm{pr}'(x)(x^2 - r^2)^{-1/2}\,dx \qquad (5.15)$$

in perfect analogy to Eq. (5.12).

Note that the main mathematical difference between (5.15) and (5.12) is the appearance of the kernel $(x^2 - r^2)^{-1/2}$ instead of $(r^2 - x^2)^{+1/2}$. Because

$(x^2 - r^2)^{-1/2}$ diverges at $x = r$, we cannot remove the derivative of pr(x) by partial integration, whereas in the case of $(r^2 - x^2)^{+1/2}$ we can do this operation and obtain (5.11). This singularity is still left from the singular kernel in the inverse Radon transformation (5.10) after the phase integration. Note that the transformations (5.12) and (5.15) are related to Abel's integral [65], and in view of this they are called *Abel transformations*, after the Norwegian mathematician Niels Abel. Equation (5.15) is also called Cormack's inversion formula [64].

The inverse Abel transformation (5.15) exhibits two remarkable features. First, the Wigner function $W(r)$ does not depend on quadrature values x inside $0 \leq x < r$. It is quite easy to understand that the quadrature distribution pr(x) does not depend on the Wigner function inside a circle of radius x in phase space

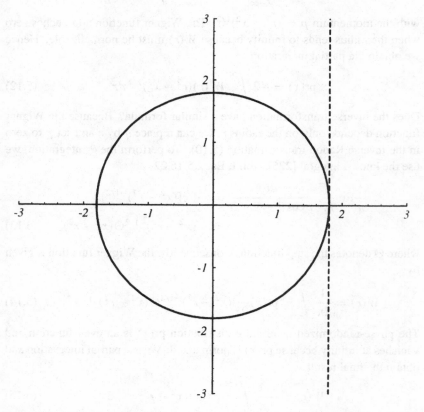

Fig. 5.3. Phase-space geometry for phase-independent Wigner functions. Only values of the Wigner function outside the depicted circle contribute to line integrals (dotted line) at the radius of this circle. Because the Wigner function $W(r)$ is uniquely determined by these integrals, $W(r)$ cannot depend on quadrature values inside a radius r as well.

[see Eq. (5.12)]. The reason is that pr(x) is given by marginal integrations of $W(r)$ along parallel lines with respect to one axis in phase space having distances x from the origin. Hence any circle of $W(r)$ values inside the radius x is not crossed and consequently does not contribute to pr(x). Because $W(r)$ is uniquely determined by pr(x), the Wigner function $W(r)$ cannot depend on quadrature values inside a radius r as well.

The second remarkable feature of Eq. (5.15) is the following: Sections of increasing probability pr(x) contribute negatively to the Wigner function, whereas sections of decreasing pr(x) contribute positively to $W(r)$. The balance of the two decides the sign of the Wigner function at radius r, indicating nonclassical behavior. To understand this property of the inverse Abel transformation (5.15), we recall again that the probability distribution pr(x) for quadratures x is given by the marginal distributions of the Wigner function. Hence negative values of $W(r)$ lead to a lack of quadrature probability, causing dips in the probability distribution pr(x). On the other hand, the Wigner function at radius r depends on pr(x) outside r only. Hence the Wigner function corresponding to the bottom of a dip in the quadrature distribution depends on the increase of pr(x). The deeper the dip is, the higher is the increase, indicating a negative, that is, a nonclassical, Wigner function.

5.1.3 Filtered back-projection algorithm

The actual numerical implementation of the inverse Radon transformation (5.10) requires an appropriate regularization of the generalized function $K(x)$, done by setting a frequency cutoff k_c in the definition (5.4) of the kernel $K(x)$. In this case we obtain the integral

$$K(x) = \frac{1}{2} \int_{-k_c}^{+k_c} |\xi| \exp(i\xi x) \, d\xi, \tag{5.16}$$

which is easily calculated to yield the result

$$K(x) = \frac{1}{x^2} [\cos(k_c x) + k_c x \sin(k_c x) - 1]. \tag{5.17}$$

However, the function $K(x)$ is not well defined at $x = 0$ (at least if numerics is concerned), because here both the numerator and the denominator tend to zero. To overcome this difficulty we expand $\cos(k_c x) + k_c x \sin(k_c x) - 1$ in powers of x and consider the first three terms

$$K(x) = \frac{k_c^2}{2} \left[1 - \frac{k_c^2 x^2}{4} + \frac{k_c^4 x^4}{72} - \cdots \right]. \tag{5.18}$$

This function is well defined at $x = 0$. A convenient point x_c to switch over
from the regularization (5.17) to the approximation (5.18) near the origin is
given by

$$|k_c x_c| = 0.1. \tag{5.19}$$

The choice of the cutoff k_c depends on the finest details of the Wigner function,
which are to be resolved without introducing rapid oscillations [brought about
by the trigonometric functions in the regularization (5.17)]. The right cutoff is
the k_c value just below the onset of the oscillations. It is advisable to adjust the
cutoff to the particular reconstruction.

With these remarks on the regularization of the kernel $K(x)$, we have sketched
the basics of the *filtered back-projection algorithm* for implementing the inverse Radon transformation (5.10). Given the cutoff k_c, the function $K(x)$ is
calculated and stored. [Figure 5.4 shows a plot of $K(x)$.] Then the Wigner
function is obtained by convolving the quadrature distributions with the kernel
$K(q \cos \theta + p \sin \theta)$ according to Eq. (5.5). The phase integration is usually
approximated by a sum with respect to the set of reference phases at which
the homodyne measurements have been performed. (In refined versions of the
algorithm, interpolation is used for the θ integration.) The x convolution can
be computed using Fourier transformation.

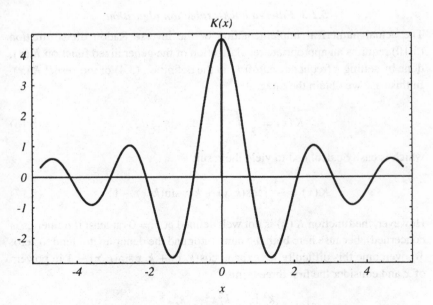

Fig. 5.4. Plot of the regularized kernel $K(x)$ with $k_c = 3$.

5.1.4 Quantum mechanics without probability amplitudes

After these ideas about the actual numerics of computerized tomography, let us turn to a more exotic topic. We have seen that quantum tomography allows us to represent the Wigner function in terms of the quadrature distributions $\mathrm{pr}(x, \theta)$, that is, in terms of observable quantities. On the other hand, we can use the overlap relation (3.20) to express the probability p_a for a transition from the state $\hat{\rho}$ to the state $\hat{\rho}_a$ in a measurement where the quantity a is observed

$$p_a = \mathrm{tr}\{\hat{\rho}\hat{\rho}_a\} = 2\pi \int_{-\infty}^{+\infty} \int_{-\infty}^{+\infty} W(q, p) W_a(q, p) \, dq \, dp. \qquad (5.20)$$

Here $W_a(q, p)$ denotes the Wigner function of $\hat{\rho}_a$. As pointed out in the very beginning, in Sections 1.3.3 and 1.3.4, this quantity is the key to quantum mechanics. Everything else follows – operators and the Hilbert-space formalism. Moreover, because we have considered mixed states (and not only pure state vectors) in their Wigner representation W and W_a, we have even left room for an imprecise knowledge of the actual pure state or of the measurement result (or of both).

In the inverse Radon transformation (5.10) the Wigner function is given in terms of the quadrature distributions. Consequently, we can certainly express the transition probability (5.20) in terms of observable quantities. Using the characteristic functions $\tilde{W}$ and $\tilde{W}_a$, we obtain from Eq. (5.10)

$$\mathrm{tr}\{\hat{\rho}\hat{\rho}_a\} = \frac{1}{2\pi} \int_{-\infty}^{+\infty} \int_{-\infty}^{+\infty} \tilde{W}(u, v) \tilde{W}_a(-u, -v) \, du \, dv \qquad (5.21)$$

$$= \frac{1}{2\pi} \int_{0}^{\pi} \int_{-\infty}^{+\infty} \widetilde{\mathrm{pr}}(\xi, \theta) \widetilde{\mathrm{pr}}_a(-\xi, \theta) |\xi| \, d\xi \, d\theta. \qquad (5.22)$$

Here $\widetilde{\mathrm{pr}}$ and $\widetilde{\mathrm{pr}}_a$ denote the Fourier-transformed quadrature distributions corresponding to $\hat{\rho}$ and $\hat{\rho}_a$, and we have used the central relation (3.9) between $\widetilde{\mathrm{pr}}$ and $\widetilde{\mathrm{pr}}_a$ and the characteristic functions. We replace $\widetilde{\mathrm{pr}}$ and $\widetilde{\mathrm{pr}}_a$ by their definition (3.5) and obtain along similar lines as in Section 5.1.1 the final result

$$\mathrm{tr}\{\hat{\rho}\hat{\rho}_a\} = -\frac{\mathcal{P}}{\pi} \int_{0}^{\pi} \int_{-\infty}^{+\infty} \int_{-\infty}^{+\infty} \frac{\mathrm{pr}(q, \theta) \mathrm{pr}_a(q_a, \theta)}{(q - q_a)^2} \, dq \, dq_a \, d\theta. \qquad (5.23)$$

This formula quantifies the transition probability from the state $\hat{\rho}$ to the state $\hat{\rho}_a$ in terms of the quadrature distributions $\mathrm{pr}(q, \theta)$ and $\mathrm{pr}_a(q_a, \theta)$.

We can consider the quantitative description of this probability to be the key axiom of quantum mechanics. Consequently, we can formulate the fundamentals of quantum physics without ever mentioning abstract Hilbert spaces and the underlying superposition principle in the first place. The quantum state is legitimately represented by the quadrature distribution $\mathrm{pr}(q, \theta)$, which is an observable quantity, and we no longer need an obscure density operator. Moreover, in classical statistics we would obtain the very formula (5.23) for a transition probability as well. To see this we interpret $W(q, p)$ and $W_a(q, p)$ as the phase-space densities for the classical states before and after the measurement, respectively. The transition probability is proportional to the joint probability $W(q, p)W_a(q, p) \, dq \, dp$ of being in the phase-space cell $dq \, dp$ integrated with respect to all q and p. (The "particle" must be localized in any of the phase-space cells before and after the measurement.) The proportionality constant is a universal quantity having the physical units of an action, say, Planck's constant $h = 2\pi\hbar$. In this book physical units are scaled such that $\hbar$ equals unity, and so we obtain the factor 2π in front of the overlap integral (5.20). Because the quadrature distributions are simply rotated marginal distributions of $W(q, p)$, they are related to the phase-space density in a classical way *by definition*. Consequently, formula (5.23) describes the transition from the state before and after the measurement in classical physics as well. Quantum mechanics has been reduced to classical statistics. But can this *really* be true?

The first and obvious objection to this reasoning is that the Wigner function might be negative. This true quantum feature is not clearly visible in the quadrature distributions, yet of course it is plainly displayed in the Wigner representation. We cannot interpret negative Wigner functions as phase-space densities, and hence the entire argument is flawed. Negative Wigner functions indicate nonclassical behavior. We may raise a second more subtle yet deeper objection. We have seen in Section 3.1.2 that many constraints are imposed on a proper Wigner function. The freedom of the system to have a certain "phase-space density" is restricted. No clear way of seeing all the constraints in the Wigner representation is possible, in general. However, all restrictions come down to the requirement that any density matrix derived from the Wigner function according to Eq. (3.27) can be seen as a statistical mixture of pure Hilbert-space vectors. So it is much more convenient to describe the freedom the system has, that is, the state, in terms of the density operator. We are back to Hilbert space as the basic structure of quantum mechanics. We see that in fact the *superposition principle* lies at the heart of what quantum mechanics is about. In addition to the statistical uncertainty in the outcomes of quantum measurements – the quantum jumps – the superposition principle provides quantum mechanics with a unique and highly nontrivial structure. The principle contains physics and not sheer logic. Finally, we remark that the expression (5.23) is simply far less beautiful than the equivalent overlap $\text{tr}\{\hat{\rho}\hat{\rho}_a\}$ of density operators. Additionally, subsequent formulas (for instance, describing the time evolution) would be even more awkward compared with the elegance of the Hilbert-space formalism. After all, beauty is a clear indication for truth.

5.2 Quantum-state sampling

We have seen how to reconstruct the Wigner function from the measured quadrature distributions by means of computerized tomography. We know that the Wigner function comprises all information we need to determine the quantum state of the investigated spatial–temporal light mode. In fact, we can simply integrate a reconstructed Wigner function with respect to a set of known functions to obtain the density matrix according to formula (3.27). Because this integration and the inverse Radon transformation (5.10) used to infer the Wigner function are both linear integral transformations, a linear expression for the density matrix in terms of the quadrature distributions must necessarily exist

$$\rho_{a'a} \equiv \langle a'|\hat{\rho}|a\rangle \tag{5.24}$$

$$= \int_0^\pi \int_{-\infty}^{+\infty} \mathrm{pr}(q, \theta) F_{a'a}(q, \theta) \, dq \, d\theta, \tag{5.25}$$

where $\{|a\rangle\}$ is an arbitrary basis. The kernel $F_{a'a}(q, \theta)$ simply combines the inverse Radon transformation (5.10) with the overlap relation (3.27) for reconstructing the density matrix.

Considered purely mathematically, the expression (5.25) is trivial and is of course equivalent to the previously discussed tomographic version of quantum-state inference. So what difference does the expression make? To see this we write formula (5.25) as a weighted average (denoted by double brackets) of the functions $F_{a'a}(q, \theta)$ with respect to the measured q and θ values

$$\rho_{a'a} = \langle\langle F_{a'a}(q, \theta)\rangle\rangle_{q,\theta}. \tag{5.26}$$

This average must not be confused with a quantum-mechanical expectation value. Formula (5.26) describes a statistical averaging with respect to the outcomes (or parameters) of quantum measurements. Only observable quantities are involved – the quadrature distributions $\mathrm{pr}(q, \theta)$ and the functions $F_{a'a}(q, \theta)$ for processing the data. Formula (5.26) means that the density matrix can be *statistically sampled* by measuring quadrature values.

The sampling may be performed "on line": Every datum (q, θ) of the series of homodyne measurements contributes individually to the density matrix. The functions $F_{a'a}(q, \theta)$ are calculated and averaged so that the matrix is gradually building up. Once statistical confidence in the sampled density matrix is sufficient, the experiment can be stopped. (A way to estimate the statistical accuracy of the sampled results will be derived in Section 5.3.4.) So we can directly monitor the buildup of the density matrix using a numerical implementation of formula (5.26). We also note that direct sampling does not rely on storing the quadrature distributions. Only the density matrix must be stored, which may save computer memory because usually this matrix is much smaller than the quadrature histograms.

Moreover, in quantum-state sampling the regularization (5.17) of the inverse Radon transformation (5.10) is avoided. Instead, the sampling functions $F_{a'a}(q, \theta)$ have absorbed the mathematical effect of the convolution with the singular kernel $-\mathcal{P}x^{-2}$. If these functions are well behaved (and they are for normalizable basis vectors), the convergence of the statistical sampling is guaranteed. The sampling method produces the "true picture" of the quantum state, whereas the tomographic technique involves some filtering. Yet paradoxically, this filtering may be a practically useful side effect because it smoothes out the statistical distribution $\mathrm{pr}(q, \theta)$ of the measured quadratures. Filtering acts like an anticipation of the quadrature histograms from a finite set of data. Because the sampling method reveals the "true picture," it is also more sensitive to statistical errors. The choice between the two data-processing techniques is mostly a matter of practical convenience and depends on the particular experimental situation.

Last but not least, we can learn a lot about the practical issues of optical-homodyne tomography by examining the mathematical properties of the sampling functions. Our estimations of the statistical errors, of the general feasibility of homodyne tomography, of the number of reference phases, and of the required quadrature bin width are all based on this analysis.

5.2.1 Pattern functions and Hilbert transformation

After these remarks on the practical implications of the simple transformation (5.25), let us examine the mathematics of the sampling functions. We would like to find analytically tale-telling and numerically convenient expressions for the functions $F_{a'a}(q, \theta)$. Probably the simplest approach to the structure of the sampling functions is to proceed as in Section 5.1.4, where we were concerned about quantum mechanics without probability amplitudes. In Eq. (5.23) we had expressed the quantum overlap $\mathrm{tr}\{\hat{\rho}\hat{\rho}_a\}$ in terms of the quadrature distributions for the quantum states $\hat{\rho}$ and $\hat{\rho}_a$. Because the density-matrix element $\rho_{a'a}$ is a quantum overlap of the density operator $\hat{\rho}$ and the projector $|a\rangle\langle a'|$ according to Eq. (5.24), we can use the same arguments as in Section 5.1.4. The only difference is that here we define the generalized "quadrature distribution"

$$\mathrm{pr}_{a'a}(x, \theta) \equiv \langle x|\hat{U}(\theta)|a\rangle\langle a'|\hat{U}^\dagger(\theta)|x\rangle \qquad (5.27)$$

of the projector $|a\rangle\langle a'|$ instead of the Hermitian operator $\hat{\rho}_a$. [Here $|x\rangle$ denotes a quadrature eigenstate, and $\hat{U}(\theta)$ is the familiar phase-shifting operator defined in Eq. (2.6).] This "distribution" might be complex-valued, but in Section 5.1.4

we have actually never used the fact that the quadrature distribution $pr_a(q, \theta)$ is real. Consequently we obtain the formula

$$\rho_{a'a} = -\frac{\mathcal{P}}{\pi} \int_0^\pi \int_{-\infty}^{+\infty} \int_{-\infty}^{+\infty} \frac{pr(q, \theta)\, pr_{a'a}(x, \theta)}{(q - x)^2}\, dx\, dq\, d\theta \qquad (5.28)$$

in complete analogy to Eq. (5.23). By comparing this expression with Eq. (5.25), we see that we can write the density matrix in terms of the sampling functions

$$F_{a'a}(q, \theta) = -\frac{\mathcal{P}}{\pi} \int_{-\infty}^{+\infty} \frac{pr_{a'a}(x, \theta)}{(q - x)^2}\, dx. \qquad (5.29)$$

Formula (5.29) shows strikingly that the functions $F_{a'a}(q, \theta)$ have indeed absorbed the tomographic kernel $-\mathcal{P}x^{-2}$. This kernel is sharply peaked, and so the sampling functions resemble the generalized quadrature distributions of the projector $|a\rangle\langle a'|$ within the accuracy of an algebraically decaying function x^{-2}. In fact, the $F_{a'a}(q, \theta)$ display just the typical pattern of the density-matrix projector $|a\rangle\langle a'|$ in the quadrature representation. They detect these patterns in the measured quadrature distributions $pr(q, \theta)$. In view of this fact, the sampling functions $F_{a'a}(q, \theta)$ are also called *pattern functions* [167].

We may express an $F_{a'a}(q, \theta)$ as the derivative

$$F_{a'a}(q, \theta) = \frac{\partial}{\partial q} G_{a'a}(q, \theta) \qquad (5.30)$$

of another function $G_{a'a}(q, \theta)$ given by the formula

$$G_{a'a}(q, \theta) = \frac{\mathcal{P}}{\pi} \int_{-\infty}^{+\infty} \frac{pr_{a'a}(x, \theta)}{q - x}\, dx. \qquad (5.31)$$

Integral transformations such as (5.31) are called *Hilbert transformations* and are well known in complex analysis and especially in the theory of Fourier transformations [264]. In optics, Hilbert transformations are most famous as the mathematical basis for the Kramers–Kronig relations [265] and in the theory of analytic signals [38]. Note that the mathematics of Hilbert transformations can be used for finding convenient expressions for the pattern functions of a given basis [169]. Here, however, we will proceed in a different way to derive our results and we will use formula (5.29) only for studying some general properties of the pattern functions.

For instance, we easily find the asymptotic behavior of the $F_{a'a}(q, \theta)$ for large q using a simple multipole expansion of the kernel

$$-(q - x)^{-2} \sim -q^{-2} - 2xq^{-3} + \cdots. \qquad (5.32)$$

For the main-diagonal projectors $|a\rangle\langle a|$ the quadrature distributions $pr_{aa}(x, \theta)$ are normalized to unity. Consequently, the pattern functions tend to

$$F_{aa}(q, \theta) \sim -\pi^{-1}q^{-2} \qquad (5.33)$$

for large q, as we easily see by inserting the first term of the multipole expansion (5.32) in Eq. (5.29). If the basis system $\{|a\rangle\}$ is orthonormal, the integral $\int_{-\infty}^{+\infty} \mathrm{pr}_{a'a}(q, \theta)\, dq = \langle a' | a \rangle$ must vanish for $a \neq a'$. In this case the $F_{a'a}(q, \theta)$ decay at least like q^{-3}. Note that the algebraic decay of the pattern functions is related to the algebraic resolution of the tomographic kernel $-\mathcal{P}x^{-2}$. It ensures that the integration (5.25) with respect to the quadrature distributions $\mathrm{pr}(q, \theta)$ converges.

Let us study two prominent examples of pattern functions, those of the Fock basis and of the coherent-state basis. Let us consider the Fock basis first. The generalized quadrature distributions of the number-state projectors $|n\rangle\langle m|$ are given by the expression

$$\mathrm{pr}_{mn}(x, \theta) = \langle x|\hat{U}(\theta)|n\rangle\langle m|\hat{U}^\dagger(\theta)|x\rangle$$
$$= \psi_m(x)\psi_n(x)\exp[\mathrm{i}(m - n)\theta]. \tag{5.34}$$

Here the $\psi_n(x)$ denote the (real) quadrature wave functions for the Fock states $|n\rangle$. We have used the fact that the phase-shifting operator $\hat{U}(\theta)$ defined in Eq. (2.6) provides a Fock state $|n\rangle$ with the phase factor $\exp(-\mathrm{i}n\theta)$. Consequently, we obtain a simple phase dependence for the pattern functions of the Fock basis

$$F_{mn}(q, \theta) = \frac{1}{\pi} f_{mn}(q)\exp[\mathrm{i}(m - n)\theta] \tag{5.35}$$

where we have introduced the *amplitude pattern functions*

$$f_{mn}(q) = -\mathcal{P}\int_{-\infty}^{+\infty} \frac{\psi_m(x)\psi_n(x)}{(q - x)^2}\, dx. \tag{5.36}$$

(Note that $\pi^{-1}f_{mn}(q)$ instead of $f_{mn}(q)$ has also been [70], [167], [169] called a pattern function. We have, however, good reasons [168] for prefering the notation (5.36) here; see Section 5.2.3.) We note that the amplitude pattern functions $f_{mn}(q)$ are even functions for even differences $m - n$ and odd functions for odd ones

$$f_{mn}(-q) = (-1)^{m-n}f_{mn}(q) \tag{5.37}$$

because the product of the quadrature wave functions in Eq. (5.36) has the same property. We also note that the quadrature probability distribution must obey the simple symmetry relation

$$\mathrm{pr}(q, \theta + \pi) = \mathrm{pr}(-q, \theta). \tag{5.38}$$

This property is an immediate consequence of the Radon transformation (5.1). After a half cycle π the harmonic-oscillator wave packet is reversed. We may use the relations (5.37) and (5.38) to extend the sampling formula (5.25) to a complete oscillation cycle

$$\rho_{mn} = \frac{1}{2\pi} \int_{-\pi}^{+\pi} \int_{-\infty}^{+\infty} \text{pr}(q, \theta) f_{mn}(q) \exp[\text{i}(m - n)\theta] \, dq \, d\theta. \qquad (5.39)$$

This extension is always possible as long as the amplitude pattern functions obey the symmetry (5.37). On the other hand, if the $f_{mn}(q)$ are symmetric (5.37), we can reduce the quantum-state sampling within the full phase interval $[-\pi, \pi]$ to the sampling (5.25) within $[0, \pi]$ only.

Another prominent case is the set of the pattern functions for the coherent-state basis. We consider again the generalized quadrature distributions of the relevant density-matrix projector $|\alpha\rangle\langle\alpha'|$ and get

$$\text{pr}_{\alpha'\alpha}(x, \theta) = \langle x|\hat{U}(\theta)|\alpha\rangle\langle\alpha'|\hat{U}^\dagger(\theta)|x\rangle$$
$$= \langle x|\alpha\exp(-\text{i}\theta)\rangle\langle\alpha'\exp(-\text{i}\theta)|x\rangle, \qquad (5.40)$$

where we have used the phase-shifter property (2.47) of coherent states. It takes only some straightforward algebra to obtain from the expression (2.60) for the wave functions $\langle x \mid \alpha \rangle$ of coherent states $|\alpha\rangle$ the result

$$\text{pr}_{\alpha'\alpha}(x, \theta) = \pi^{-1/2} \exp\left[-(x - x_\theta)^2 - \frac{1}{2}|\alpha - \alpha'|^2\right] \qquad (5.41)$$

with the parameter

$$x_\theta = 2^{-1/2}[\alpha\exp(-\text{i}\theta) + \alpha'^*\exp(+\text{i}\theta)]. \qquad (5.42)$$

Because the quadrature distribution for the vacuum state is just $\pi^{-1/2}\exp(-x^2)$ [see Eq. (2.33)], we find immediately from formula (5.29) that the pattern functions of the coherent-state basis are given by

$$F_{\alpha'\alpha}(q, \theta) = \frac{1}{\pi} f_{00}(q - x_\theta) \exp\left(-\frac{1}{2}|\alpha - \alpha'|^2\right). \qquad (5.43)$$

Considered theoretically, coherent states are displaced vacuums. So it is not surprising that the pattern functions for coherent states are basically displaced vacuum pattern functions $\pi^{-1} f_{00}(q)$. The displacement parameter x_θ, however, is complex for the off-diagonal matrix elements where α differs from α'.

5.2.2 Random phase and photon statistics

Most quantum-optical experiments involving homodyne detection are essentially interferometers. A master laser generates light. One part of this original light is guided to the apparatus for producing the signal, that is, light having certain interesting properties, like nonclassical features, for instance. Another part serves as the local oscillator beam in the homodyne measurement. Finally, both light fields are recombined at the homodyne detector. This interferometric

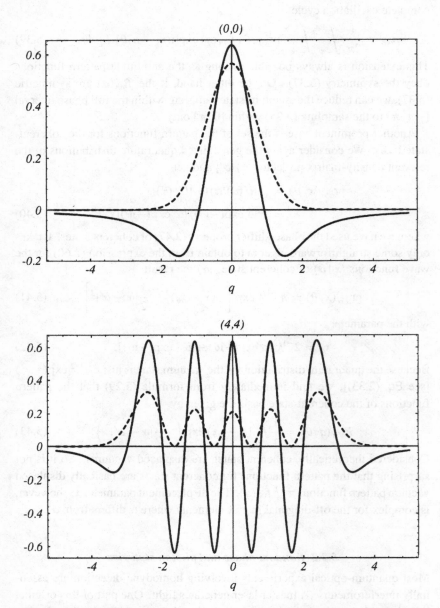

Fig. 5.5. Amplitude pattern functions $f_{mn}(q)$ for the Fock basis (solid lines) versus the product of the wave functions $\psi_n(q)$ and $\psi_m(q)$ (dashed lines). Diagonal case. Top: $n = m = 0$ (vacuum component), bottom: $n = m = 4$ (four-photon Fock state).

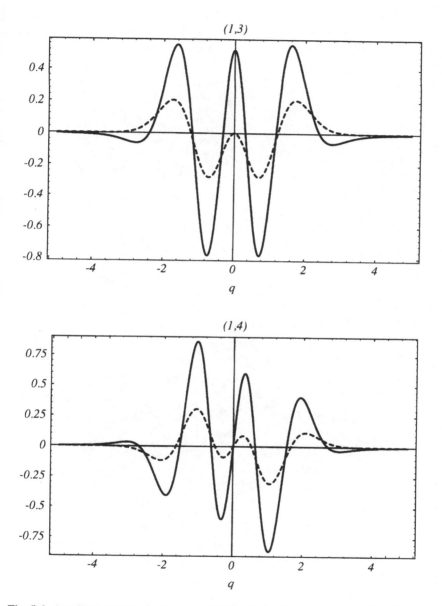

Fig. 5.6. Amplitude pattern functions $f_{mn}(q)$ for the Fock basis (solid lines) versus the product of the wave functions $\psi_n(q)$ and $\psi_m(q)$ (dashed lines). Off-diagonal case. Top: $n = 1, m = 3$, bottom: $n = 1, m = 4$.

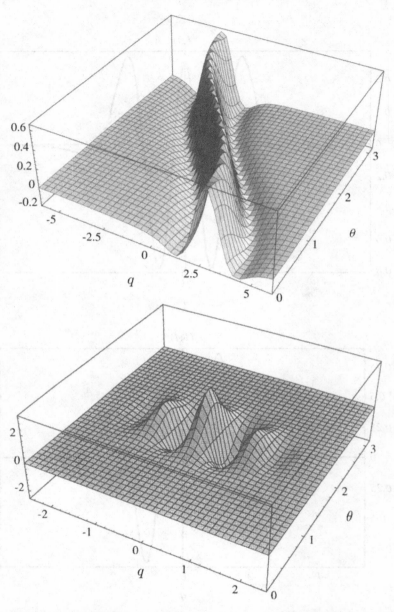

Fig. 5.7. Pattern functions $F_{\alpha'\alpha}(q,\theta)$ for the coherent-state representation. We see that the diagonal pattern function $F_{\alpha\alpha}(q,\theta)$ (top) with $\alpha = 2$ follows the track of the harmonic-oscillator evolution brought about by the phase shift θ. The real part of the off-diagonal pattern function (bottom) with $\alpha' = 2$ and $\alpha = -2$ resembles the typical Schrödinger-cat oscillations, and in this way detects quantum coherences in the measured quadrature distributions.

setup guarantees that the relative phase between the local oscillator and the signal is controlled. However, in other experiments the two light fields may originate from different master lasers. In this case the relative phase is drifting and cannot be controlled anymore. Under ideal circumstances this drift is uniform, so that the phase difference between the local oscillator and the signal is random in repeated homodyne measurements. It is also possible that the generation of the signal involves phase-incoherent processes, so that the relative phase between signal and LO is randomized even in an interferometric setup. In all these cases we have lost the phase information of the signal, yet we may still measure the intensity properties. In quantum optics the most detailed intensity information is provided by the photon statistics

$$p_n = \rho_{nn}. \tag{5.44}$$

We assume that here the homodyne detector measures the phase-randomized quadrature distribution

$$\mathrm{pr}(q) = \frac{1}{2\pi} \int_{-\pi}^{+\pi} \mathrm{pr}(q,\theta)\, d\theta. \tag{5.45}$$

We use the sampling formula (5.39) to represent the photon statistics

$$p_n = \int_{-\infty}^{+\infty} \mathrm{pr}(q) M_n(q)\, dq \tag{5.46}$$

in terms of the phase-randomized quadrature distribution $\mathrm{pr}(q)$. The functions

$$M_n(q) = f_{nn}(q) \tag{5.47}$$

sample the photon statistics from phase-randomized homodyne data. This indirect method for measuring the photon-number distribution in phase-randomized homodyne detection was first proposed and experimentally demonstrated by Munroe et al. [193]. Because of the high efficiency and single-photon resolution of balanced homodyne detection, this technique is today's best way of obtaining detailed intensity information about light on the quantum level. Moreover, time-resolved measurements of the photon statistics can be performed using local-oscillator pulses of controlled spatial–temporal shapes. Finally, we note that phase randomization eliminates artificial features in the reconstructed photon statistics brought about by a finite number of LO phases.

5.2.3 *A theorem on the Schrödinger equation*

What are the sampling functions? How do they behave, and how do we calculate them? Surprisingly, the answer to all these questions is not restricted to the case of harmonic oscillators [168], for example, to the light modes considered

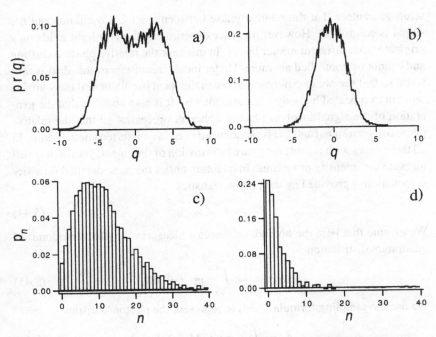

Fig. 5.8. Phase-randomized quadrature histograms (a, b) and reconstructed photon-number distributions (c, d). Reproduced from the first experimental demonstration [193] of the photon-number sampling via homodyne detection.

in this book. We may generalize the idea of quantum-state sampling to one-dimensional wave packets moving in arbitrary potentials, using a theorem [168] on the one-dimensional Schrödinger equation. Although this generalization goes beyond the scope of this book, we prefer not to withhold this connection to a broader physical situation. The idea could be useful for state measurements of anharmonic molecules, of wave packets in semiconductor quantum wells, or of electronic wave packets in atoms, to name just a few examples.

Let us first summarize what we have done so far and express the result in a different way. Using homodyne detection the quadrature distribution

$$\mathrm{pr}(q, \theta) = \sum_{\mu\nu} \rho_{\mu\nu} \psi_\mu(q) \psi_\nu(q) \exp[i(\nu - \mu)\theta] \qquad (5.48)$$

can be measured. In Eq. (5.48) we have expressed $\mathrm{pr}(q, \theta)$ in terms of the (real) Fock-state wave functions $\psi_\nu(q)$ and the density matrix $\rho_{\mu\nu}$ in Fock representation. This formula, together with Eq. (5.39), shows that we can divide the sampling of the density matrix into two steps. First, the measured quadrature distribution $\mathrm{pr}(q, \theta)$ is Fourier-transformed with respect to θ, and we obtain from Eq. (5.48)

$$\widetilde{\mathrm{pr}}(q, m - n) \equiv \frac{1}{2\pi} \int_{-\pi}^{+\pi} \mathrm{pr}(q, \theta) \exp[\mathrm{i}(m - n)\theta] \, d\theta \qquad (5.49)$$

$$= \sum_{\mu\nu}{}' \rho_{\mu\nu} \psi_\mu(q) \psi_\nu(q) \qquad (5.50)$$

where the summation $\sum'$ is restricted by the constraint

$$\mu - \nu = m - n. \qquad (5.51)$$

The second step is the integration with respect to the amplitude pattern functions $f_{mn}(q)$. This operation projects the density-matrix elements ρ_{mn} out of the Fourier-transformed quadrature distributions

$$\rho_{mn} = \int_{-\infty}^{+\infty} \widetilde{\mathrm{pr}}(q, m - n) f_{mn}(q) \, dq. \qquad (5.52)$$

A sufficient and necessary condition for this property is that the $f_{mn}(q)$ are orthonormal

$$\int_{-\infty}^{+\infty} \psi_\mu(q) \psi_\nu(q) f_{mn}(q) \, dq = \delta_{\mu m} \delta_{\nu n} \qquad (5.53)$$

to products of the wave functions $\psi_\mu(q)$ and $\psi_\nu(q)$, given the constraint (5.51). Obviously, the amplitude pattern functions (5.36) satisfy this orthogonality relation (5.53). Because the property (5.53) is all we need to sample the density matrix, we may regard Eq. (5.53) as an alternative definition of the $f_{mn}(q)$.

How can we generalize these ideas to wave packets moving in arbitrary potentials? We simply identify the time t of the motion with the phase shift θ and the position x with the quadrature amplitude q, and we consider an arbitrary potential instead of the harmonic $q^2/2$. We express the position probability distribution $\mathrm{pr}(x, t)$ of the moving wave packet

$$\mathrm{pr}(x, t) = \sum_{\mu\nu} \rho_{\mu\nu} \psi_\mu(x) \psi_\nu(x) \exp[\mathrm{i}(\omega_\nu - \omega_\mu)t] \qquad (5.54)$$

in terms of the density matrix $\rho_{\mu\nu}$ in energy representation, the stationary wave functions $\psi_n(x)$, and the eigenfrequencies (energies) ω_n. We assume that only the discrete part of the spectrum is excited. Because the potential $U(x)$ is real, the eigenfunctions $\psi_n(x)$ can be chosen to be real as well. Because none of the discrete levels is degenerate in one dimension [145, Section 21] $\psi_n(x)$ is the only normalized solution of the stationary Schrödinger equation with eigenfrequency ω_n

$$\left[-\frac{1}{2} \frac{\partial^2}{\partial x^2} + U(x) \right] \psi_n(x) = \omega_n \psi_n(x). \qquad (5.55)$$

We suppose that the position probability distribution $\text{pr}(x, t)$ can be measured at all times t. (In general, this measurement might be a highly nontrivial experimental challenge. Thanks to homodyne detection, the measurement of $\text{pr}(q, \theta)$ is relatively easy for the electromagnetic oscillator in quantum optics.)

How is the density matrix ρ_{mn} inferred from the observed motion $\text{pr}(x, t)$ of the wave packet? We may proceed again in two steps. First, we perform a temporal Fourier transformation

$$\widetilde{\text{pr}}(x, \omega_m - \omega_n) \equiv \frac{1}{T} \int_{-T/2}^{+T/2} \text{pr}(x, t) \exp[i(\omega_m - \omega_n)t]\, dt \qquad (5.56)$$

$$= \sum_{\mu\nu}{}' \rho_{\mu\nu} \psi_\mu(x) \psi_\nu(x) \qquad (5.57)$$

to distinguish the density-matrix elements $\rho_{\mu\nu}$ that oscillate at the difference frequency

$$\omega_\mu - \omega_\nu = \omega_m - \omega_n. \qquad (5.58)$$

For the harmonic oscillator the sampling time T is one 2π cycle. In the general case, however, T must be sufficiently large so that it includes many cycles. The second step of our procedure is again the integration

$$\rho_{mn} = \int_{-\infty}^{+\infty} \widetilde{\text{pr}}(x, \omega_m - \omega_n) f_{mn}(x)\, dx \qquad (5.59)$$

with respect to a set of spatial sampling functions $f_{mn}(x)$. If they are orthonormal (5.53) to products of the wave functions $\psi_\mu(x)\psi_\nu(x)$ with the frequency constraint (5.58), then the density matrix ρ_{mn} is indeed inferred.

Surprisingly, the sampling functions $f_{mn}(x)$ turn out to be quite simple [168]. They are just the first derivatives of products of *regular and irregular wave functions*

$$f_{mn}(x) = \frac{\partial}{\partial x}[\psi_m(x)\varphi_n(x)]. \qquad (5.60)$$

What are irregular wave functions? Any linear differential equation of second order, like the Schrödinger equation (5.55), must have two linearly independent solutions for a given frequency ω_n. One is the regular wave function ψ_n: It is normalizable for certain eigenfrequencies ω_n leading to the quantization of energy. The other fundamental solution φ_n attached to ω_n is called *irregular*. Because the stationary states are nondegenerate [145, Section 21], φ_n cannot be normalizable, as ψ_n is, and must be discarded as a physical state. Note that irregular wave functions have nevertheless been used in scattering theory. See for instance Refs. [98] and [145, Section 138]. We prove in Appendix 2 that if the Wronskian W_n of the two solutions ψ_n and φ_n equals 2,

$$W_n \equiv \psi_n \varphi_n' - \psi_n' \varphi_n = 2, \qquad (5.61)$$

the function $f_{mn}(x)$ of Eq. (5.60) is orthonormal (5.53) to the product of the wave functions $\psi_\mu(x)\,\psi_\nu(x)$, given the constraint (5.58). This theorem is the one we use. [We note that here a prime symbolizes the first spatial derivative. We also note that the Wronskian (5.61) is always a spatial constant, that is, $W_n' = 0$, as is easily verified using the stationary Schrödinger equation (5.55). Any two solutions ψ_n and φ_n of (5.55) produce a certain value of the Wronskian (5.61)]. The Wronskian condition (5.61) shows that φ_n must be irregular indeed, for otherwise the Schrödinger equation (5.55) would have two linearly independent and normalizable solutions, and so ω_n would be degenerate. Apart from the condition (5.61) the irregular wave function φ_n can be freely chosen from all solutions of the stationary Schrödinger equation (5.55) with the eigenfrequency ω_n. Moreover, we may exchange the regular and irregular part in the expression (5.60). According to our theorem, $f_{nm}(x)$ is orthonormal to the product $\psi_\nu(q)\psi_\mu(q)$, which is of course the same as $\psi_\mu(q)\psi_\nu(q)$. So $f_{nm}(x)$ as well as $f_{mn}(x)$ satisfies the orthonormality condition (5.53). The density matrix is still statistically sampled. All these ambiguities of the spatial sampling functions leave enough room for choosing the numerically most convenient forms.

Finally, we comprise the essence of quantum-state sampling in a single formula [171]. We consider the density operator $\hat{\rho}$ in a general basis

$$\rho_{a'a} = \sum_{mn} \langle a' \mid m \rangle \rho_{mn} \langle n \mid a \rangle. \qquad (5.62)$$

The time-dependent regular wave functions $\psi_a(x, t)$ for the basis states $\mid a \rangle$ are given by

$$\psi_a(x, t) = \sum_n \langle n \mid a \rangle \psi_n(x) \exp(-i\omega_n t). \qquad (5.63)$$

They are of course solutions of a time-dependent Schrödinger equation with the potential $U(x)$ and the initial condition $\psi_a(x) = \langle x \mid a \rangle$. We define irregular wave functions $\varphi_a(x, t)$ as

$$\varphi_a(x, t) \equiv \sum_n \langle n \mid a \rangle \varphi_n(x) \exp(-i\omega_n t). \qquad (5.64)$$

Using these expressions, we recombine the separately considered sampling steps (5.56) and (5.59) in the final formula

$$\rho_{a'a} = \left\langle\!\!\left\langle \frac{\partial}{\partial x} [\psi_{a'}^*(x, t)\varphi_a(x, t)] \right\rangle\!\!\right\rangle_{x,t}. \qquad (5.65)$$

The double brackets denote an average with respect to the experimentally measured (x, t) data. The observation of the moving wave packet reveals the quantum state at $t = 0$. Of course, we need to know the dynamical law of motion, that is, the potential $U(x)$ in the Schrödinger equation (5.55) to calculate the required regular and irregular wave functions. Formula (5.65) shows how to sample the density matrix from observation of the position x evolving in time t.

Let us close this section with a few remarks on our special case, the harmonic oscillator. Although we have found the sampling formula (5.65) for the density matrix $\rho_{a'a}$ in an arbitrary basis, $\partial[\psi_{a'}^*(q, \theta)\varphi_a(q, \theta)]/\partial q$ is not necessarily equal to the pattern function $F_{a'a}(q, \theta)$ defined in (5.29), although both serve the same purpose. The reason is the general nonuniqueness of the sampling functions. According to our theorem an amplitude pattern function $f_{mn}(q)$ in the Fock representation would be given by $\partial[\psi_m(q)\varphi_n(q)]/\partial q$. Note that we can always choose an odd irregular wave function $\varphi_n(q)$ for even n and vice versa (but not a φ_n having the same parity as the regular ψ_n). In this way we can construct amplitude pattern functions with the symmetry (5.37). As already mentioned, this symmetry is required for reducing the sampling (5.39) within a complete cycle $[-\pi, \pi]$ to our familiar sampling (5.25) within the phase interval $[0, \pi]$.

5.2.4 Irregular wave functions

How do we calculate the irregular wave function for Fock states? What properties do they have? How do they look? Quite early in this book, in Eq. (2.41), we introduced the irregular wave function $\varphi_0(q)$ for the vacuum state as an alternative yet not normalizable solution of the Schrödinger equation. The function $\varphi_0(q)$ is odd, whereas the vacuum state has an even, regular wave function $\psi_0(q)$ given by the Gaussian $\pi^{-1/4} \exp(-q^2/2)$. We normalize the irregular wave function φ_0 of Eq. (2.41) in such a way that the Wronskian

$$W_0 = \psi_0\varphi_0' - \psi_0'\varphi_0 = c\,2^{1/2}\pi^{1/4} \tag{5.66}$$

equals 2 and obtain

$$\varphi_0(q) = \pi^{3/4} \exp\left(-\frac{q^2}{2}\right) \text{erfi}(q). \tag{5.67}$$

As in Eq. (2.41), erfi denotes the imaginary error function defined by the integral (2.42). Consequently, the amplitude pattern function $f_{00}(q)$ for the vacuum state is given by the expression

$$f_{00}(q) = 2[1 - \pi^{1/2}q \exp(-q^2)\, \text{erfi}(q)]. \tag{5.68}$$

Inspired by the formula (2.35) for the regular wave functions of excited states, we define the irregular ones as "excitations" of the irregular vacuum

$$\varphi_n(q) = \frac{\hat{a}^{\dagger n}}{\sqrt{n!}} \varphi_0(q). \tag{5.69}$$

As an immediate consequence of this definition, the irregular wave functions are related to each other by excitation steps

$$\hat{a}^\dagger \varphi_n = \sqrt{n+1} \varphi_{n+1}. \tag{5.70}$$

In the Schrödinger representation the annihilation and creation operators read

$$\hat{a} = \frac{1}{\sqrt{2}} \left(q + \frac{\partial}{\partial q} \right), \qquad \hat{a}^\dagger = \frac{1}{\sqrt{2}} \left(q - \frac{\partial}{\partial q} \right). \tag{5.71}$$

Because of their definition (5.69) and the structure (5.67) of the irregular vacuum, the irregular wave functions for Fock states involve only certain polynomials and $\exp(-q^2/2)\mathrm{erfi}(q)$ terms. Our definition (5.69) guarantees that $\varphi_n(q)$ is a solution of the stationary Schrödinger equation

$$\left[-\frac{1}{2} \frac{\partial^2}{\partial q^2} + \frac{q^2}{2} \right] \varphi_n = \left(\hat{a}^\dagger \hat{a} + \frac{1}{2} \right) \varphi_n = \left(n + \frac{1}{2} \right) \varphi_n. \tag{5.72}$$

To verify this relation we note that if φ_n is a solution of (5.72), then φ_{n+1} satisfies $\hat{a}^\dagger \hat{a} \varphi_{n+1} = (n+1)\varphi_{n+1}$, as is easily seen using formula (5.70) and the bosonic commutation relation $[\hat{a}, \hat{a}^\dagger] = 1$. The irregular vacuum $\varphi_0(q)$ obeys $\hat{a}^\dagger \hat{a} \varphi_0 = 0$ by definitions (2.38) and (2.39). This implies that the irregular wave function φ_n for an excited state $|n\rangle$ must be indeed a solution of the stationary Schrödinger equation (5.72).

However, are the φ_n the right solutions? We must check whether all Wronskians (5.61) equal 2, as is the case for the vacuum. Are the Wronskians W_n conserved during the excitation (5.69) of the irregular vacuum (5.67)? From the relation (5.70) using the Schrödinger representation (5.71) we immediately obtain

$$\varphi_{n+1} = \frac{1}{\sqrt{2n+2}} (q\varphi_n - \varphi_n'). \tag{5.73}$$

We differentiate this formula and use the Schrödinger equation (5.72) to find a relation for the derivative

$$\varphi_{n+1}' = \frac{1}{\sqrt{2n+2}} [(2n+2)\varphi_n + q\varphi_n' - q^2\varphi_n]. \tag{5.74}$$

Obviously, the rules (5.73) and (5.74) are also valid for the regular wave functions ψ_n and their derivatives ψ_n', because we have used only the excitation formula (5.69) and the Schrödinger equation (5.72), that is, common properties of φ_n and ψ_n. Applying these relations, we easily verify that the Wronskian (5.61) is indeed conserved

$$W_{n+1} = W_n. \tag{5.75}$$

Consequently, the Wronskians W_n must equal 2 for all irregular wave functions defined by Eq. (5.69). Additionally, our definition provides the φ_n with the desired symmetry properties for reducing the sampling (5.39) to the phase interval $[0, \pi]$. To see this we note that the operator $\hat{a}^\dagger = 2^{-1/2}(q - \partial/\partial q)$ is odd (it changes sign if q is replaced by $-q$). As a consequence, φ_{n+1} must be even if φ_n is odd and vice versa. Because the irregular vacuum φ_0 is odd whereas ψ_0 is even, all irregular solutions φ_n must have exactly the opposite parity of the regular wave functions ψ_n. We have seen that the excitations (5.69) of the irregular vacuum (5.67) qualify as appropriate irregular wave functions.

What happens if we "annihilate" an irregular excitation φ_n? We know that a regular wave function ψ_n produces $\sqrt{n}\psi_{n-1}$ when the annihilation operator $\hat{a}$ is applied to it. See Eq. (2.26). To understand the action of $\hat{a}$ in the irregular case, we may use the following trick: As a consequence of the bosonic commutation relation $[\hat{a}, \hat{a}^\dagger] = 1$, we get $\hat{a}\hat{a}^{\dagger n} = \hat{a}^\dagger \hat{a}\hat{a}^{\dagger(n-1)} + \hat{a}^{\dagger(n-1)}$ and obtain by repeated applications of this rule $\hat{a}\,\hat{a}^{\dagger n} = \hat{a}^{\dagger(n-1)}(\hat{a}^\dagger \hat{a} + n)$. This implies that $\hat{a}\varphi_n$ gives $\hat{a}^{\dagger(n-1)}(\hat{a}^\dagger \hat{a} + n)\,\varphi_0/\sqrt{n!}$ according to the definition (5.69) of the irregular wave functions φ_n. Because φ_0 is the irregular vacuum with $\hat{a}^\dagger \hat{a}\varphi_0 = 0$, we obtain

$$\hat{a}\varphi_n = \sqrt{n}\varphi_{n-1}. \qquad (5.76)$$

As in the regular case, the annihilation operator $\hat{a}$ "annihilates" the irregular wave functions. The only difference is that the rule (5.76) is valid only for $n > 0$ because the annihilation of the irregular vacuum produces instead of zero the function φ_{-1} by definition (2.38). We combine the creation and annihilation rules (5.70) and (5.76) using the Schrödinger representation (5.71) and obtain an important recurrence relation of the irregular wave functions for $n > 0$

$$\sqrt{n+1}\varphi_{n+1} + \sqrt{n}\varphi_{n-1} = \sqrt{2}q\varphi_n. \qquad (5.77)$$

The same relation holds in the regular case for $n \geq 0$. We see that the algebra of the irregular wave functions is very similar indeed to the regular algebra of the Fock states.

To compare the regular and the irregular wave functions further we use the semiclassical approximation developed in Appendix 1. It turns out that in the classically allowed region $|q| < r_n$ bounded by the Bohr–Sommerfeld radius r_n of Eq. (3.75), both solutions of the stationary Schrödinger equation (5.72) oscillate like standing waves

$$\psi_n \sim \sqrt{\frac{2}{\pi}}p_n^{-1/2}\cos\left(S_n + \frac{\pi}{4}\right), \qquad (5.78)$$

$$\varphi_n \sim \sqrt{2\pi}\,p_n^{-1/2}\sin\left(S_n + \frac{\pi}{4}\right). \qquad (5.79)$$

Here p_n abbreviates the semiclassical momentum (it must not be confused with the photon-number probability, which is unfortunately denoted by the same symbol in the literature). Using the parameterization

$$q = r_n \cos t_n \qquad (5.80)$$

for the position q, the semiclassical momentum p_n is given by

$$p_n = r_n \sin t_n. \qquad (5.81)$$

The parameter t_m plays simply the role of the semiclassical oscillation time. The quantity

$$S_n \equiv \int_{r_n}^{q} p_n(x) \, dx = \frac{r_n^2}{4}[\sin(2t_n) - 2t_n] \qquad (5.82)$$

denotes the time-independent part of the classical action. The prefactor $\sqrt{2/\pi}$ in Eq. (5.78) is determined in Appendix 1, whereas the factor $\sqrt{2\pi}$ in Eq. (5.79) does its duty in the Wronskian condition (5.61). We see from the semiclassical formulas (5.78) and (5.79) that the regular and the irregular wave functions are oscillating out of phase in the classically allowed region. Both are standing Schrödinger waves, that is, interfering running waves $p_n^{-1/2} \exp(iS_n)$ of the action S_n, yet in the classically forbidden zone $|q| > r_n$ this interference becomes destructive for the regular wave functions and constructive for the irregular solutions. The regular wave functions decay for large q, whereas the irregular solutions grow (and hence they are not normalizable).

How fast do the irregular wave functions grow? First we find the asymptotic behavior of the irregular vacuum solution $\varphi_0(q)$. We use the fact that erfi$'(q)$ gives $2\pi^{-1/2} \exp(q^2)$ by definition (2.42) and that the function $[q^{-1} \exp(q^2)]'$ tends to $2 \exp(q^2)$ in leading order for large q. In this way we obtain

$$\varphi_0(q) \sim \pi^{1/4} q^{-1} \exp\left(\frac{q^2}{2}\right). \qquad (5.83)$$

Consequently, $\varphi_n(q)$ tends to

$$\varphi_n(q) \sim \left(\frac{n!\sqrt{\pi}}{2^n}\right)^{1/2} q^{-n-1} \exp\left(\frac{q^2}{2}\right) \qquad (5.84)$$

for large q because this formula satisfies definition (5.69) with an initial $\varphi_0(q)$ given by the asymptotic expression (5.83). Similarly, we obtain from the regular vacuum (2.33) using the relation (2.36) in leading order

$$\psi_n(q) \sim \left(\frac{2^n}{n!\sqrt{\pi}}\right)^{1/2} q^n \exp\left(-\frac{q^2}{2}\right). \qquad (5.85)$$

In this way we see that although the irregular wave functions grow exponentially for large arguments q, the amplitude pattern functions (5.60) tend to

$$f_{mn}(q) \sim \left(\frac{n!}{m!} 2^{m-n}\right)^{1/2} (m - n - 1)q^{m-n-2}. \qquad (5.86)$$

In particular, the amplitude pattern functions $f_{nn}(q)$ for the diagonal elements of the density matrix decay like $-q^{-2}$ for large q. This is consistent with the general rule (5.33). On the other hand, the off–diagonal amplitude pattern functions $f_{mn}(q)$ decay algebraically for $n \geq m$ and grow for $n < m - 1$. However, we may take advantage of the general ambiguity of the sampling functions $f_{mn}(q)$ to define

$$f_{mn}(q) = \begin{cases} \partial[\psi_m(q)\varphi_n(q)]/\partial q & \text{for } n \geq m \\ \partial[\psi_n(q)\varphi_m(q)]/\partial q & \text{for } n < m. \end{cases} \quad (5.87)$$

The so-defined $f_{mn}(q)$ are appropriate amplitude pattern functions, and they decay always algebraically for large q. This property is numerically convenient in evaluations of sampling integrals such as (5.52). Note that an analysis [169] of the $\psi_m(q)\varphi_n(q)$ for complex q proves that the expression (5.87) is indeed the unique solution of the Hilbert transformation (5.36). So we have found some convenient mathematical expressions for the amplitude pattern functions $f_{mn}(q)$ of the Fock basis. In particular we have developed the annihilation-and-creation formalism for the irregular wave functions.

5.2.5 *Numerical recipes*

After these mathematical excursions we have everything on hand to sketch the actual numerical procedure for calculating the amplitude pattern functions. We use the formula (5.73) for the irregular wave functions and the equivalent expression for the regular ones to express the derivatives $\varphi'_n(q)$ and $\psi'_n(q)$ in terms of the corresponding wave functions. In this way we obtain the final result

$$f_{mn}(q) = 2q\psi_m(q)\varphi_n(q) - \sqrt{2(m + 1)}\psi_{m+1}(q)\varphi_n(q)$$
$$- \sqrt{2(n + 1)}\psi_m(q)\varphi_{n+1}(q) \quad (5.88)$$

for the amplitude pattern functions with $n \geq m$. Otherwise we can use the symmetry

$$f_{mn}(q) = f_{nm}(q). \quad (5.89)$$

We see from formula (5.88) that the pattern functions $f_{mn}(q)$ of the whole density *matrix* depend on just two *vectors*, $\psi_m(q)$ and $\varphi_n(q)$. This simplifies significantly the numerical effort needed. Moreover, the vectors ψ_m and φ_n are easily computed using recurrence relations such as (5.77). For the regular wave functions ψ_n we recommend applying the relation

$$\psi_m = \frac{1}{\sqrt{m}}[\sqrt{2}q\psi_{m-1} - \sqrt{m - 1}\psi_{m-2}] \quad (5.90)$$

with the start values

$$\psi_0 = \pi^{-1/4} \exp\left(-\frac{q^2}{2}\right), \tag{5.91}$$

$$\psi_1 = \pi^{-1/4}\sqrt{2}q \exp\left(-\frac{q^2}{2}\right). \tag{5.92}$$

However, using this kind of forward recursion for the irregular wave functions φ_n as well is not advisable because it depends too critically on the accuracy of the initial values (which involve the imaginary error function). Instead, we may turn the tables and apply a backward recursion starting from the semiclassical solution for high quantum numbers. This procedure defines a sequence of functions that stably converges to the exact irregular wave functions. For a density matrix with maximal quantum number M, we must cover at least the oscillating part in the range of quadrature values q by the classically allowed region for the initial irregular wave function, which is then semiclassically approximated. The quantum number $4M$ is a safe choice for this. The classically allowed region for $4M$ excluding the Bohr–Sommerfeld band is given roughly by

$$|q| < r_{4M} - (2r_{4M})^{-1/3} \tag{5.93}$$

with the Bohr–Sommerfeld radius

$$r_n = \sqrt{2n+1}. \tag{5.94}$$

This is motivated by the scaling of the irregular wave functions in the Bohr–Sommerfeld band. See Appendix 1. Inside the region (5.93) we recommend using the backward recursion

$$\varphi_n = \frac{1}{\sqrt{n+1}}[\sqrt{2}q\varphi_{n+1} - \sqrt{n+2}\varphi_{n+2}] \tag{5.95}$$

with initial values for $n = 4M, 4M-1$ given by the semiclassical approximation (5.79)

$$\varphi_n = \left(\frac{2\pi}{r_n \sin t_n}\right)^{1/2} \sin\left[\frac{r_n^2}{4}(\sin(2t_n) - 2t_n) + \frac{\pi}{4}\right]. \tag{5.96}$$

The parameter t_n equals

$$t_n = \arccos\left(q/r_n\right). \tag{5.97}$$

Outside the safe region (5.93) we recommend using the asymptotic expression (5.84) for φ_n. It is easily implemented by the forward recursion

$$\varphi_n = \left(\frac{n}{2}\right)^{1/2} q^{-1}\varphi_{n-1} \tag{5.98}$$

with the start value

$$\varphi_0 = \pi^{1/4}q^{-1}\exp\left(\frac{q^2}{2}\right). \tag{5.99}$$

We note that the accuracy of this numerical method was tested [169] by computing the integral $\int_{-\infty}^{+\infty}\psi_k(q)^2 f_{nn}(q)\,dq$, which should equal δ_{kn} according to Eq. (5.53). It turned out that the accuracy was always about 10^{-5} or better.

Using this procedure for computing the pattern functions $f_{mn}(q)$, the density matrix ρ_{mn} in Fock basis can be sampled from homodyne measurements as a statistical average

$$\rho_{mn} = \left\langle\left\langle \frac{1}{\pi}f_{mn}(q)\exp[i(m-n)\theta]\right\rangle\right\rangle_{q,\theta} \tag{5.100}$$

with respect to quadrature amplitudes q and local-oscillator phases θ. In this way the quantum state can be sampled *in real time* during data collection. Only the relevant current information, that is, the matrix ρ_{mn}, is stored, which saves time and computer memory.

5.2.6 Quantum and classical tomography

Of course, once we have determined the density matrix, we can calculate the Wigner function as well. Let us derive a numerically efficient algorithm for this procedure. We assume that the density matrix is given in the Fock basis, that is, by the expression

$$\hat{\rho} = \sum_{m,n=0}^{M}\rho_{mn}|m\rangle\langle n|, \tag{5.101}$$

with the highest quantum number M, which implies that ρ_{mn} can be truncated to a good approximation. To calculate the Wigner function $W(q,p)$, we insert the expansion (5.101) in Wigner's formula (3.17) and obtain

$$W(q,p) = \sum_{m,n=0}^{M}\rho_{mn}W_{mn}(q,p) \tag{5.102}$$

with

$$W_{mn}(q,p) = \frac{1}{\pi}\int_{-\infty}^{+\infty}\exp(2ipx)\langle q-x\,|\,m\rangle\langle n\,|\,q+x\rangle\,dx \tag{5.103}$$

$$= \frac{(-1)^m}{\pi}\int_{-\infty}^{+\infty}\exp(2ipx)\langle x-q\,|\,m\rangle\langle n\,|\,x+q\rangle\,dx.$$

To obtain the last line we have used the fact that the Schrödinger wave functions are even for even-number Fock states and odd for odd-number states. We compare this expression with formula (3.15) and see immediately that the Wigner function $W_{mn}(q, p)$ is essentially equivalent to the characteristic function, that is,

$$W_{mn}(q, p) = \frac{(-1)^m}{\pi} \tilde{W}_{mn}(-2p, 2q). \quad (5.104)$$

We use the key formula (3.12) for the characteristic function and get

$$W_{mn}(q, p) = \frac{(-1)^m}{\pi} \text{tr}\{|m\rangle\langle n| \exp(2ip\hat{q} - 2iq\hat{p})\}$$

$$= \frac{(-1)^m}{\pi} \langle n| \exp(2ip\hat{q} - 2iq\hat{p})|m\rangle$$

$$= \frac{(-1)^m}{\pi} \langle n|\hat{D}(2\alpha)|m\rangle \quad (5.105)$$

according to Eq. (2.54) with $\alpha = 2^{-1/2}(q + ip)$. Calculating the Fock representation of the displacement operator remains. For this step we express $\hat{D}$ in terms of annihilation and creation operators [see Eq. (2.48)], employ the Baker–Hausdorff formula (2.62), and expand the exponentials $\exp(2\alpha\hat{a}^\dagger)$ and $\exp(-2\alpha^*\hat{a})$ in Taylor series. We use the property

$$\hat{a}^\nu|n\rangle = \left[\frac{n!}{(n - \nu)!}\right]|n - \nu\rangle \quad (5.106)$$

of the Fock states [a simple consequence of Eq. (2.35)] and obtain

$$\langle n|\hat{D}(2\alpha)|m\rangle = \exp(-2|\alpha|^2) \sum_{\nu=0}^{n} \sum_{\mu=0}^{m} (2\alpha)^\nu (-2\alpha^*)^\mu$$

$$\times \frac{1}{\nu!\mu!} \left[\frac{n!m!}{(n - \nu)!(m - \mu)!}\right]^{1/2}$$

$$\times \langle n - \nu \, | \, m - \mu \rangle. \quad (5.107)$$

Because Fock states are orthonormal, we can reduce the double sum to a single sum and in particular to a certain polynomial in $|\alpha|^2$. In fact, we see from the explicit representation [89, Vol. II, Eq. 10.12.(7)] of the Laguerre polynomials L_n^k that the Fock representation of the displacement operator is simply

$$\langle n|\hat{D}(2\alpha)|m\rangle = \left(\frac{n!}{m!}\right)^{1/2} \exp\left(-2|\alpha|^2\right)(-2\alpha^*)^{m-n} L_n^{m-n}(4|\alpha|^2) \quad (5.108)$$

for $m \geq n$ and

$$\langle n|\hat{D}(2\alpha)|m\rangle = \langle m|\hat{D}(-2\alpha)|n\rangle^* \quad (5.109)$$

for $m < n$. Now we are well prepared to formulate an efficient algorithm for calculating the Wigner function $W(q, p)$, given the density matrix ρ_{mn} in the

Fock representation. We introduce polar coordinates r and φ in phase space so that q and p are given by

$$q = r \cos \varphi, \qquad p = r \sin \varphi \qquad (5.110)$$

and, consequently, α is equal to $2^{-1/2} r \exp(i\varphi)$. We use the expansion (5.101) and the formula (5.105) with the results (5.108) and (5.109), abbreviate $m - n$ by k, and obtain, finally,

$$W(q, p) = \sum_{k=-M}^{M} w(r, k) \exp(-ik\varphi) \qquad (5.111)$$

with

$$w(r, k) = \begin{cases} \sum_{n=0}^{M-k} w_n(r, k) \rho_{n+k,n} & \text{for } k \geq 0 \\ w(r, -k)^* & \text{for } k < 0 \end{cases} \qquad (5.112)$$

and

$$w_n(r, k) = \frac{1}{\pi} (-1)^n \left[\frac{n!}{(n+k)!} \right]^{1/2} \exp(-r^2)(r\sqrt{2})^k L_n^k(2r^2). \qquad (5.113)$$

These are indeed numerically convenient expressions because we can calculate $w_n(r, k)$ by the recurrence relation

$$w_n(r, k) = \frac{1}{\sqrt{n(n+k)}} [(2r^2 + 1 - k - 2n) w_{n-1}(r, k)$$

$$- \sqrt{(n-1)(n-1+k)} w_{n-2}(r, k)], \qquad (5.114)$$

which is easily derived from the relation [89, Vol. II, Eq. 10.12(8)] of the Laguerre polynomials. The initial values of the recurrence (5.114) are

$$w_0(r, k) = \frac{1}{\pi} (k!)^{-1/2} (r\sqrt{2})^k \exp(-r^2) \qquad (5.115)$$

(with $w_0(0, 0) = \pi^{-1}$) and

$$w_{-1}(r, k) = 0. \qquad (5.116)$$

We see that the $w_n(r, k)$ functions are easily calculated en passant during the summation (5.112). The algorithm is stable and fast (especially if the square roots of the required integers are stored before starting the reconstruction). No filtering is needed, provided of course that the density matrix can be indeed truncated to a good approximation.

Finally, we note that the combination of the sampling (5.100) with the sketched reconstruction algorithm can also be applied in classical tomography, where the goal is to reconstruct a two-dimensional image from a set of one-dimensional projections. The overall procedure is mathematically equivalent to the inverse Radon transformation. Although we calculate a "density matrix" ρ_{mn} as an intermediate step, this matrix serves only for storing the Fourier components of the classical projections $\mathrm{pr}(q, \theta)$ in a convenient way. The physical interpretation of classical and quantum tomography is, however, much different. In classical statistical physics the phase-space density $W(q, p)$ describes the freedom a system may have, that is, the state, whereas in quantum mechanics the density matrix is the most natural form to express quantum states (remember the discussion of "quantum mechanics without probability amplitudes" in Section 5.1.4). To put this distinction into mathematical terms, in classical physics the phase-space density is nonnegative (and the formally derived "density matrix" might be negative), whereas in quantum mechanics the density operator has nonnegative eigenvalues (and the Wigner function might be negative). However, this distinction does not matter in the numerical procedure. We note that instead of the filtering required in the usual numerical implementation of the inverse Radon transformation (filtered back projection) (see Section 5.1.3), we have used the truncation of the matrix ρ_{mn}. Our method has the advantage that we can employ the mathematical background of quantum mechanics to shed light on some practical issues of classical tomography. In particular, we derive in Section 5.3.2 how many reference phases are sufficient for an *exact* reconstruction if the "density matrix" can be truncated, and we estimate the error if this truncation is not possible. Borrowing ideas from quantum mechanics, we have found an alternative way of doing and understanding classical tomography.

5.3 How precisely can we measure quantum states?

So far we have assumed that the quadrature distribution $\mathrm{pr}(q, \theta)$ can be perfectly measured. This is, however, only approximately the case in real-world experiments, for many possible reasons. Let us focus on some universal issues that do not depend much on the particular experimental setup. What are these points? The accuracy of the quadrature value q may be limited because of *detection losses* described by the overall efficiency η. Usually, quadratures are measured with respect to a *finite number of reference phases* θ only. Additionally, in order to obtain a quadrature histogram, the range of q is divided into small bins and the bin width δq limits the *quadrature resolution*. Finally, the number N of experimental runs is finite for a given reference phase θ, that is, the available statistical ensemble is limited. This limitation gives rise to *statistical errors* in the reconstructed density matrix. All these points are important for the understanding of realistic quantum-state reconstructions. Let us consider them separately to understand what influences they have.

5.3.1 Detection losses

In Section 4.2.4 we arrived at a simple model for an inefficient homodyne detector. We can replace the homodyne apparatus by an ideal quadrature detector with a fictitious beam splitter in front of it. In this way we separate the inefficiencies from the principal character of the detection process. The beam splitter accounts for the detection losses – it transmits a fraction η of the total intensity, whereas a fraction $(1 - \eta)$ of all incident photons are ignored on average. Here η quantifies the *overall detection efficiency*, which comprises all kinds of losses. More precisely,

$$\eta = \eta_M \cdot \eta_{QE} \cdot \eta_L, \tag{5.117}$$

where η_M denotes the degree of mode matching given by Eq. (4.90), η_{QE} quantifies the efficiency of the photodetectors, and η_L accounts for experimental losses. (Usually, mode mismatch is responsible for the lion's share of overall detection losses in homodyne measurements.) Although the fictitious beam splitter transmits on average a fraction of η from every incoming photon, each photon is not split to fractions. Instead, the photons are selected at random from the stream of incident light as if they were classical particles. (The probability, however, of detecting one photon is given by η.) The randomness in the detection successes and failures causes additional statistical noise in the homodyne measurement. This noise is formally described by a vacuum entering the apparatus via the second input port of the fictitious beam splitter. Because of the detection noise the measured quadrature distribution is smoothed according to Eq. (4.98). If we reconstruct the Wigner function $W^{(r)}(q, p)$ from smoothed quadrature distributions, this function is accordingly smoothed compared to the Wigner function $W(q, p)$ of the original signal. On the other hand, if faithfully reconstructed, $W^{(r)}(q, p)$ is of course the true Wigner function of the signal after partial absorption [255]. We know from our simple absorber theory developed in Section 4.1.4, and in particular from Eqs. (4.49) and (4.50) that after absorption we obtain an s-parameterized quasiprobability distribution

$$W^{(r)}(q, p) = \eta^{-1} W(\eta^{-1/2}q, \eta^{-1/2}p; 1 - \eta^{-1}) \tag{5.118}$$

instead of the Wigner function $W(q, p)$ of the initial signal. The quadratures q and p are scaled according to the loss of intensity, and the negative parameter $s = 1 - \eta^{-1}$ describes the smoothing of the original Wigner function, that is, the extra noise involved. In the case of fifty percent overall efficiency, the Q function is reconstructed. Counting every second photon smoothes out any potential negative regions the Wigner function might have. This shows again that a high overall detection efficiency is required to *see* nonclassical effects.

According to Section 4.1.4 the reconstructed density matrix in Fock representation $\langle m|\hat{\rho}^{(r)}|n\rangle$ has suffered from the *generalized Bernoulli transformation* (4.58). In particular, the photon-number distribution $p_n = \langle n|\hat{\rho}|n\rangle$ has been smoothed in the obtained distribution $p_n^{(r)} = \langle n|\hat{\rho}^{(r)}|n\rangle$ according to Eq. (4.59). For small losses and low photon numbers we obtain in the first order in $(1-\eta)$

$$p_n^{(r)} \approx p_n + (n+1)(1-\eta)p_{n+1}. \tag{5.119}$$

This equation shows that the next neighbor higher on the energy ladder, p_{n+1}, contributes to $p_n^{(r)}$ and so smoothes out the photon-number distribution. In particular, the pairing (2.93) and (2.94) of photons in a squeezed vacuum is easily destroyed by detection losses and therefore difficult to measure. Only recently [240] has this effect been clearly observed (see the discussion at the end of Section 2.3 and in particular Fig. 2.5).

Can we compensate numerically for the effect of losses after the measurement has been performed? We have already mentioned in Sections 3.2 and 3.3 that it is very difficult to eliminate the smoothing from an s-parameterized quasiprobability distribution (5.118) for retrieving the original Wigner function, if this distribution (5.118) is numerically or experimentally given. Can we perform the required deconvolution directly on the density matrix $\langle m|\hat{\rho}|n\rangle$? In principle we could easily *invert* the generalized Bernoulli transformation (4.58). To see this we use our simple absorber model. At the beginning of Section 4.1.4 we postulated that after absorption the P function of the field is scaled (4.46) by the factor $\eta^{-1/2}$. To invert this operation we simply rescale the argument of P by $\eta^{+1/2}$. We imagine the density matrix (4.58) as being derived from the P function via the overlap relation (3.69), for instance. Because the function (4.58) is analytic in $\eta^{1/2}$ we simply replace η by η^{-1} to yield the original density matrix $\langle m|\hat{\rho}|n\rangle$ from the rescaled P function. In this way we obtain the *generalized inverse Bernoulli transformation* [136], [148]

$$\langle m|\hat{\rho}|n\rangle = \eta^{-(m+n)/2} \sum_{k=0}^{\infty} \langle m+k|\hat{\rho}^{(r)}|n+k\rangle$$
$$\times \left[\binom{m+k}{m}\binom{n+k}{n}\right]^{1/2}(1-\eta^{-1})^k. \tag{5.120}$$

For weakly excited states, this way of numerical loss compensation is superior to a straightforward deconvolution of the reconstructed quasiprobability distribution. The reason is simple: Phase-space functions contain redundant information in the case of low quantum numbers, and much effort is therefore wasted by reconstructing these superfluous structures. However, the sketched method still leads to problems in practical applications. The sign of $(1-\eta^{-1})^k$ oscillates in the course of the summation (5.120). So the differences of matrix

elements matter in the calculation of $\langle m|\hat{\rho}|n\rangle$. This implies that the recon-
structed density matrix $\langle m|\hat{\rho}^{(r)}|n\rangle$ must be very precisely known to avoid artifi-
cial features such as negative photon-number probabilities, for instance. If the
overall efficiency is below fifty percent, the convergence of the series (5.120)
is questionable [136]. (We could, however, compensate for the losses in mul-
tiple runs of the inverse Bernoulli transformation where each run takes less
than fifty percent. We could also employ a further analytic continuation of the
density matrix as a function of $\eta^{1/2}$ [115].) We note that the efficiency η must
be precisely known to avoid overshooting in the numerical loss compensation.
Finally, almost needless to say, the best way of enhancing the efficiency of a
homodyne measurement is to minimize detection losses in the first place.

5.3.2 *Finite number of reference phases*

In principle, phase-space tomography or quantum-state sampling requires the
measurement of quadratures for *all phases* θ within the interval $[0, \pi]$. In
practice, however, only a finite number d of reference phases can be employed.
In most cases the phases θ_k are equidistant, that is,

$$\theta_k = \frac{\pi}{d}k \qquad (5.121)$$

where the integer k runs from 0 to $d - 1$. What is the effect of using d equidistant reference phases [172]? The reconstructed density matrix $\rho_{mn}^{(r)}$ in Fock
representation is given by the expression

$$\rho_{mn}^{(r)} = \frac{1}{d}\sum_{k=0}^{d-1}\int_{-\infty}^{+\infty} \mathrm{pr}(q, \theta_k)\, f_{mn}(q)\exp[\mathrm{i}(m - n)\theta_k]\, dq, \qquad (5.122)$$

irrespective of the method we use for reconstruction (sampling or phase-space
tomography). We use the symmetries (5.37) and (5.38) to extend the summation
(5.122) from $-d$ to $d - 1$. We represent the quadrature distribution $\mathrm{pr}(q, \theta)$
in terms of the Fock wave functions $\psi_\nu(q)$ [see Eq. (5.48)] and introduce the
overlap integrals

$$G_{\mu\nu}^{mn} \equiv \int_{-\infty}^{+\infty} \psi_\mu(q)\psi_\nu(q)\, f_{mn}(q)\, dq. \qquad (5.123)$$

In this way we obtain from Eq. (5.122)

$$\rho_{mn}^{(r)} = \sum_{\mu\nu} \rho_{\mu\nu} G_{\mu\nu}^{mn}\delta(\mu - \nu - m + n; 2d) \qquad (5.124)$$

with the abbreviation

$$\delta(\nu; 2d) = \frac{1}{2d}\sum_{k=-d}^{d-1} \exp(\mathrm{i}\nu\theta_k) \qquad (5.125)$$

and θ_k given by Eq. (5.121). The quantity $\delta(\nu; 2d)$ turns out to act as a *modular Kronecker symbol*, that is, it yields unity if ν is divisible by $2d$ and otherwise zero

$$\delta(\nu; 2d) = \begin{cases} 1 & \text{for } \nu = 0 \ (\text{mod } 2d) \\ 0 & \text{for } \nu \neq 0 \ (\text{mod } 2d) \end{cases}. \tag{5.126}$$

To see this we abbreviate $\exp(i\pi \nu/d)$ by z so that $\exp(i\nu\theta_k)$ equals z^k, and we use the relation $(1-z) \sum_{k=-d}^{d-1} z^k = z^{-d} - z^d = 0$. Therefore, if z differs from unity, that is, if ν is not divisible by $2d$, the sum $\sum_{k=-d}^{d-1} z^k$ must yield zero. On the other hand, if ν is divisible by $2d$, then z equals unity and the sum $\sum_{k=-d}^{d-1} z^k$ gives just $2d$. This proves the property (5.126) of the symbol $\delta(\nu; 2d)$.

Consequently, the reconstructed density matrix $\rho_{mn}^{(r)}$ is given by the formula

$$\rho_{mn}^{(r)} = \sum_{l=-\infty}^{\infty} \sum_{\mu\nu}^{l} \rho_{\mu\nu} G_{\mu\nu}^{mn} \tag{5.127}$$

where the summation $\sum^l$ is restricted by

$$\mu - \nu = m - n + 2dl \tag{5.128}$$

with integers l. Instead of obeying the exact constraint (5.51), the photon numbers $\mu - \nu$ must equal $m - n$ up to multiples of $2d$ only. This property defines the difference between the exact density matrix ρ_{mn} and the reconstructed $\rho_{mn}^{(r)}$, caused by the finite number d of equidistant reference phases. In other words, using the phase resolution π/d we cannot discriminate between phase oscillations in the quadrature distribution $\text{pr}(q, \theta)$ having difference frequencies of multiples of $2d$, a phenomenon familiar from aliasing [224]. Unfortunately, the exact number constraint (5.51) is essential for the orthonormality of the amplitude pattern functions $f_{mn}(q)$ with respect to the products $\psi_\mu(q)\psi_\nu(q)$ of the wave functions. Only in this case

$$G_{\mu\nu}^{mn} = \delta_{\mu m}\delta_{\nu n} \tag{5.129}$$

holds so that the density matrix ρ_{mn} is correctly reconstructed in $\rho_{mn}^{(r)}$. When does this matter and when not? Suppose that no quantum numbers higher than $M = d - 1$ are excited, so that the density matrix is truncated. The dimension of the state is d. In this case the deviations (5.128) from the exact number constraint (5.51) are never probed. And, of course, this is also true if the dimension of the state is smaller. So the dimension of the system defines the minimal number of reference phases needed to reconstruct the density matrix. If we can estimate a priori that no quantum numbers higher than $M = d - 1$ will occur in our state of interest, then we know with certainty that d reference phases are sufficient. Otherwise, the reconstructed density matrix may exhibit artificial features (such as negative eigenvalues).

How do we estimate the reconstruction error if we cannot truncate the actual density matrix? How do we quantify the confidence in the inferred state? Let us assume that the reconstructed $\rho_{mn}^{(r)}$ agrees already sufficiently well with the true density matrix ρ_{mn}, so that we can treat the error

$$\epsilon_{mn} \equiv \left| \rho_{mn}^{(r)} - \rho_{mn} \right| = \left| \sum_{l \neq 0} \sum_{\mu\nu}{}^{l} \rho_{\mu\nu} G_{\mu\nu}^{mn} \right| \tag{5.130}$$

as a mere correction. We replace the actual density matrix $\rho_{\mu\nu}$ in the exact error formula (5.130) by the reconstructed matrix $\rho_{\mu\nu}^{(r)}$ and obtain the error estimation

$$\epsilon_{mn} \approx \left| \sum_{l \neq 0} \sum_{\mu\nu}{}^{l} \rho_{\mu\nu}^{(r)} G_{\mu\nu}^{mn} \right|. \tag{5.131}$$

Sums of $\rho_{\mu\nu}^{(r)} G_{\mu\nu}^{mn}$ with respect to a set of lines parallel to the diagonal of the reconstructed density matrix quantify the error. In particular, off-diagonal elements are required to estimate the error of the photon-number distribution ρ_{nn}. The greater the number of phases used, the larger is the distance between these lines and the smaller is the error. (The matrix elements $\rho_{\mu\nu}^{(r)}$ decay for $\mu, \nu > d$ if the reconstruction is already sufficiently precise, so that ϵ_{mn} can be considered a correction.) We note that in the numerical procedure [169] for the pattern functions, the $G_{\mu\nu}^{mn}$ coefficients (5.123) can be easily calculated on the side.

In addition to our simple rule "number of phases = dimension of the density matrix in Fock representation," we have developed a convenient method for calculating the reconstruction error caused by a finite number of phases. The method allows an iterative state reconstruction: If the error is too large, the homodyne measurements should be repeated at the intermediate phases $(\theta_k + \theta_{k+1})/2$ until the accuracy is satisfying. Figure 5.9 shows how few phases are already sufficient to reconstruct the photon-number distribution of a Schrödinger-cat state, despite the complicated shape of the "cat" in the Wigner representation; see Fig. 3.6. Figure 5.9 thus illustrates the accuracy of our simple estimation.

5.3.3 *Quadrature resolution*

How do we determine the quadrature histograms? Usually, the range of q is divided into narrow bins with equal width δq. The number of measured quadrature values falling into the bins is counted for each bin. In this way the histogram is obtained. Also, in quantum-state sampling the quadrature range is usually divided into bins. They should not be too small in order to avoid statistical errors.

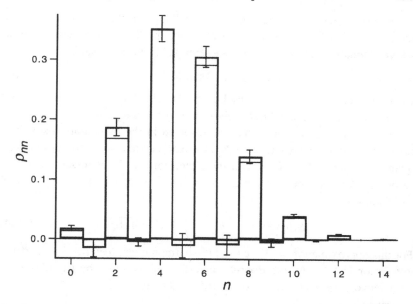

Fig. 5.9. Numerical simulation [172] to illustrate our error estimation. We used six reference phases to reconstruct the photon statistics of a Schrödinger-cat state $|\psi\rangle \propto (|\alpha\rangle + |-\alpha\rangle)$ with $\alpha^2 = 5$. The reconstructed values (thick line) agree with the actual values (thin line) within the error bars obtained from Eq. (5.131). Although we have used far fewer phases than recommended according to the rule "number of phases = effective dimension of the density matrix in Fock representation," the precision of the reconstruction is good.

(It takes a large number of samples to fill narrow bins.) On the other hand, if the resolution is not fine enough, then certain details of the quadrature distribution leave the mesh undetected. How can we estimate the required quadrature bin width?

Certainly, a detailed quadrature resolution is most relevant for relatively high quantum numbers, because the $\mathrm{pr}(q, \theta)$ oscillate most rapidly for highly excited states. We may use the WKB theory for the regular and irregular wave function (Eq. (5.78) and Eq. (5.79)) to find a semiclassical approximation for the amplitude pattern functions. We use the relation (5.82) and neglect changes in the slowly varying semiclassical momenta p_n to obtain from our result (5.87) the semiclassical formula

$$f_{nm}(q) \sim 2(p_n p_m)^{-1/2} \left[p_m \cos\left(S_n + \frac{\pi}{4} \right) \cos\left(S_m + \frac{\pi}{4} \right) \right.$$

$$\left. - p_n \sin\left(S_n + \frac{\pi}{4} \right) \sin\left(S_m + \frac{\pi}{4} \right) \right]. \qquad (5.132)$$

The amplitude pattern functions are oscillating in the classically allowed region. These oscillations are most rapid for the diagonal pattern functions where n equals m. Here we obtain the remarkably simple result

$$f_{nn}(q) \sim -2\sin[2S_n(q)], \tag{5.133}$$

where the action $S_n(q)$ is given by Eq. (5.82). The diagonal pattern functions oscillate between -2 and 2 in the classically allowed region.

Oscillations such as these must be resolved in the quadrature histograms $\mathrm{pr}(q, \theta)$ in order to reconstruct the density matrix up to a cutoff for the maximal quantum number $M = d - 1$ (for the dimension d). To estimate the needed resolution δq, we linearize the action S_n at $q = 0$, where the most rapid oscillations occur,

$$S_n(q) \approx -\frac{\pi}{2}\left(n + \frac{1}{2}\right) + \sqrt{2n + 1}\, q. \tag{5.134}$$

For this approximation we have used the harmonic parameterization (5.80) in expression (5.82) with the Bohr–Sommerfeld radius given by Eq. (5.94). A complete oscillation cycle q_M for the maximal quantum number $M = d - 1$ is described by the relation

$$2S_M(q_M) - 2S_M(0) = 2\pi. \tag{5.135}$$

We obtain for the period of the pattern functions the value

$$q_M = \frac{\pi}{\sqrt{2M + 1}}. \tag{5.136}$$

In order to resolve a period q_M the needed resolution δq should be better than $q_M/2$. Consequently, the reconstruction of the density matrix up to a maximal quantum number $M = d - 1$ (for dimension d) requires a quadrature bin width δq narrower than $q_M/2$, with q_M given by the simple formula (5.136). Note that this result is also valid for the phase-space tomography based on the inverse Radon transformation, because quantum-state sampling is mathematically equivalent to it.

5.3.4 Statistical errors

Quantum tomography is a quantitatively reliable experimental technique only if we can estimate the statistical errors of the determined quantum state or of the photon statistics. How much confidence in the density matrix do we have, given a finite set of experimental data? Again, the sampling method provides the answer. We use simply the standard recipe for estimating the propagation of experimental errors in the limit of large but not infinitely many samples. Both the quadratures q and the phase θ are divided into discrete,

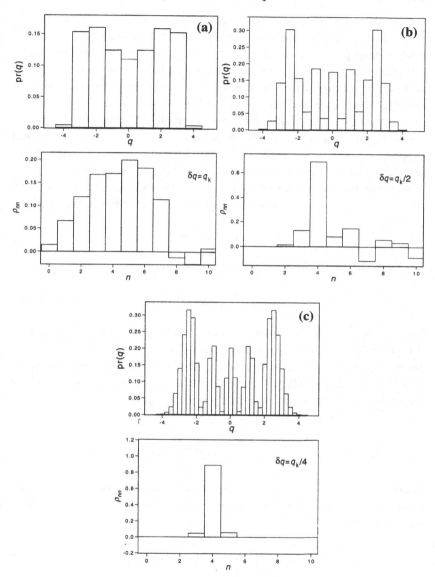

Fig. 5.10. Effect of undersampling the quadrature distribution seen in the reconstructed photon statistics [169]. We consider a Fock state $|k\rangle$ with $k = 4$ and $q_k = q_4$. Because in (a) the bin width δq equals q_4, the photon statistics are clearly undersampled. The bin width $\delta q = q_4/2$ in (b) lies on the borderline between undersampling and adequate sampling, which is seen by the peak at $n = 4$ in the reconstructed photon-number distribution. In (c) the bin width δq equals $q_4/4$, and so the photon statistics for the Fock state $|k = 4\rangle$ are almost perfectly reproduced.

narrow bins considered here as differentials dq and $d\theta$ for simplicity. The result of an individual quantum measurement is unpredictable. (Quantum-measurement processes are ultimately the best random-number generators.) The rate, however, at which the quadrature values are falling into a bin $dq\,d\theta$ is governed by the expectation value $\mathrm{pr}(q,\theta)$. So homodyne detection is a classic example of a Poisson counting process. For a finite number of samples, the experimentally measured histogram $\mathrm{pr}(q,\theta)$ at phase θ is itself a statistically fluctuating quantity. The histogram approaches the ideal quantum-mechanical expectation value $\mathrm{pr}_{id}(q,\theta)$ for a large number N of experimental runs. More precisely, $\mathrm{pr}(q,\theta)$ fluctuates statistically around $\mathrm{pr}_{id}(q,\theta)$ with a variance $\sigma_{q,\theta}^2$ equal to the mean divided by the number of samples $N(\theta)/\pi$ per phase interval $d\theta$, because the measurement process is Poissonian. In this way we obtain

$$\sigma_{q,\theta}^2 = \frac{\pi}{N(\theta)}\mathrm{pr}_{id}(q,\theta). \tag{5.137}$$

The total number of samples, or the number of single measurements, is

$$N = \pi^{-1}\int_0^\pi N(\theta)\,d\theta. \tag{5.138}$$

The measured quadrature values in each bin $dq\,d\theta$ fluctuate independently. (Strictly speaking, the normalization of the quadrature histograms sets one external constraint on the statistical fluctuations. Here we neglect this constraint, which implies that we slightly overestimate the statistical uncertainties.) The total variance σ_{mn}^2 for the real part of the reconstructed density-matrix element ρ_{mn} of Eq. (5.100) is given by

$$\sigma_{mn}^2 = \frac{1}{\pi^2}\int_0^\pi\int_{-\infty}^{+\infty}\sigma_{q,\theta}^2 f_{mn}^2(q)\cos^2[(n-m)\theta]\,dq\,d\theta. \tag{5.139}$$

In order to estimate σ_{mn}^2 directly from the measured data, we approximate the quantum-mechanical expectation value $\mathrm{pr}_{id}(q,\theta)$ by the measured histogram $\mathrm{pr}(q,\theta)$ and obtain

$$\sigma_{mn}^2 = \left\langle\left\langle\frac{1}{\pi N(\theta)}f_{mn}^2(q)\cos^2[(n-m)\theta]\right\rangle\right\rangle_{q,\theta}, \tag{5.140}$$

where the bracket means again an average over the experimental data. If all histograms $\mathrm{pr}(q,\theta)$ at the phases θ contain the same number N of samples per phase interval, then the variance σ_{mn}^2 is given by

$$\sigma_{mn}^2 = \frac{1}{\pi N}\left\langle\left\langle f_{mn}^2(q)\cos^2[(n-m)\theta]\right\rangle\right\rangle_{q,\theta}. \tag{5.141}$$

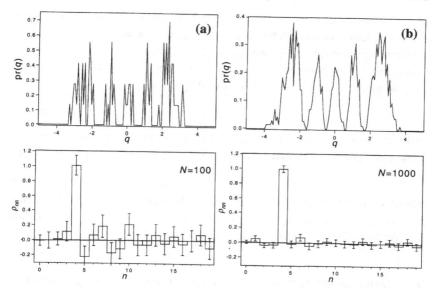

Fig. 5.11. Estimation of statistical errors in the photon-number distribution (obtained from phase-averaged quadratures) for a Fock state $|k\rangle$ with $k = 4$ [169]. The quadrature distribution was generated by a Poisson process with $N = 100$ samples in (a) and $N = 1000$ samples in (b), and the photon statistics were reconstructed (bottom). The error bars σ_{nn} obtained by our method [Eq. (5.141)] describe well the actual statistical errors. (Because the error bars are standard deviations, or square roots of the variances, they cover about sixty percent of the statistical fluctuations.)

In the same way we find the variance ς_{mn}^2 for the imaginary part of the reconstructed density-matrix element ρ_{mn}

$$\varsigma_{mn}^2 = \left\langle\!\!\left\langle \frac{1}{\pi N(\theta)} f_{mn}^2(q) \sin^2[(n-m)\theta] \right\rangle\!\!\right\rangle_{q,\theta} \tag{5.142}$$

$$= \frac{1}{\pi N} \left\langle\!\!\left\langle f_{mn}^2(q) \sin^2[(n-m)\theta] \right\rangle\!\!\right\rangle_{q,\theta} \tag{5.143}$$

for equal number of samples per phase interval $d\theta$. We see from the expressions (5.140) and (5.142) that the statistical confidence can be estimated while sampling the density matrix, meaning that experimentalists can easily estimate how many data are sufficient while taking the data, that is, they can decide when to stop the experiment.

In addition, one can also assess whether homodyne detection is feasible at all for photon-statistics or state reconstruction, *before* trying the experiment. If the number of measurements exceeds the resources of the laboratory, then there is no point in using this technique. The required number of samples depends of course on the state. However, we can apply the semiclassical approximation for the pattern functions to find a simple estimate for which very little prior information is needed. We showed in Eq. (5.133) that the main-diagonal pattern functions oscillate between -2 and $+2$. Consequently, the variance σ_{nn} of the photon-number distribution ρ_{nn} is bounded by

$$\sigma_{nn}^2 \le \frac{4}{N} \qquad (5.144)$$

using formula (5.141) and the normalization of the quadrature histograms. For a feasible experiment the statistical error must not exceed the measured quantity

$$\sigma_{nn} < \rho_{nn}, \qquad (5.145)$$

where σ_{nn} is the standard deviation, that is, the square root of σ_{nn}^2. This estimation implies that the required minimal number of samples N_{min} scales with the inverse of the photon-number probability squared

$$N_{min} \sim 4 \times \rho_{nn}^{-2}. \qquad (5.146)$$

So without knowing *a priori* the state, yet anticipating the order of magnitude of ρ_{nn}, we can estimate the required number of measurements. This simple estimation shows how difficult it is to reconstruct the photon statistics or, more generally, the density matrix for very weak signals when using homodyne detection. Because for a weak field the signal is covered by the shot noise of the LO, the extraction of ρ_{mn} requires a significant number of samples. For example, for a thermal state with mean photon number 0.01 the measurement of $\rho_{3,3}(\approx 10^{-6})$ would require about $N_{min} \sim 4 \times 10^{12}$ samples, a very large figure. For states with low occupation numbers, the price to be paid for the high efficiency and single-photon resolution of optical homodyne tomography may be a large number of experimental samples, needed for achieving statistical confidence.

5.4 Further reading

The theory of classical tomography is examined in the books by G.T. Herman [113] and F. Natterer [194], for instance. Recent results are considered in the paper [91] and the review article [92] by A. Faridani.

For more information about the first pioneering optical-homodyne-tomography experiment, see the article [256] by D.T. Smithey, M. Beck, J. Cooper, M.G. Raymer, and A. Faridani. Note that the idea of quantum tomography can also be applied to physical systems other than spatial–temporal light modes. In particular, T.J. Dunn, I.A. Walmsley, and S. Mukamel [82] reported the first tomographic quantum-state reconstruction of molecular vibrations (molecular emission tomography). M.G. Raymer, M. Beck, and D.F. McAlister [227] proposed the use of phase-space tomography to infer the quantum state of de-Broglie waves (or the two-point correlation function of classical optical fields; see also the experimental demonstration [188] by D.F. McAlister, M. Beck, L. Clarke, A. Mayer, and M.G. Raymer). For a detailed numerical simulation see the paper [126] by U. Janicke and M. Wilkens. Note that the underlying Wigner rotation is related to the fractional Fourier transformation, as was shown by A.W. Lohmann (176).

G.M. D'Ariano, C. Macchiavello, and M.B.A. Paris [68], [69] and H. Kühn, D.-G. Welsch, and W. Vogel (140) proposed the use of direct sampling instead of tomographic state reconstruction based on the inverse Radon transformation. We note that the compensation of losses could also be performed on-line during the sampling process using enhanced pattern functions [70], [167]. Because any kind of deconvolution is numerically delicate, we would, however, recommend separating the sampling process from the loss-compensation. (To obtain a clear picture first before going to retouch it is always a good idea.) The final version [169] of the sampling method for the Fock basis was the fruit of some mathematical effort; see References [70], [167]–[169], [233]. In particular, Th. Richter [233] was the first who realized that the amplitude pattern functions (for the diagonal density-matrix elements) are just derivatives of regular and irregular wave functions for the harmonic oscillator. Th. Richter and A. Wünsche [234] generalized this result to the reconstruction of the population numbers from the time-averaged motion of one-dimensional wave packets in arbitrary potentials. That the observation of moving wave packets reveals the complete quantum state was noticed in Ref. [168].

For the application of quantum tomography in classical tomographic reconstruction, see also the paper [71] by G.M. D'Ariano, C. Macchiavello, and M.B.A. Paris. The idea of quantum mechanics without probability distributions was put forward by W.K. Wootters [297] for finite-dimensional systems.

6

Simultaneous measurement of position
and momentum

6.1 Prologue

Quantum systems manage to combine classically contradicting features – the wave *and* the particle aspects, for instance. Other features they may comprise are mutually exclusive because of quantum mechanics only. According to Heisenberg's uncertainty principle, for instance, the position q and the momentum p cannot be observed both simultaneously *and* precisely. On the other hand, position and momentum define the very state of a mechanical system in classical physics (or, more generally, the statistical distribution of q and p describes an ensemble of mechanical objects). And, of course, there is no obstacle, in principle, in classical mechanics to measuring position and momentum simultaneously to an arbitrary degree of precision. In quantum mechanics these *canonically conjugate* quantities obey the commutation relation

$$i[\hat{p}, \hat{q}] = i(\hat{p}\hat{q} - \hat{q}\hat{p}) = 1 \qquad (6.1)$$

with $\hbar$ set to unity. (The relation (6.1) was discovered by Born after Heisenberg's first flash of insight into the magnificent structure of quantum mechanics in 1925 [110]. As Born declared [37], "I shall never forget the thrill I experienced when I succeeded in condensing Heisenberg's ideas on quantum conditions in the mysterious equation $\hat{p}\hat{q} - \hat{q}\hat{p} = h/2\pi i$.") According to Robertson's classic proof [236], the commutation rule (6.1) implies the uncertainty relation

$$\Delta q \, \Delta p \geq \frac{1}{2} \qquad (6.2)$$

for the standard deviations of q and p from their mean values ($\Delta^2 q \equiv \langle q^2 \rangle - \langle q \rangle^2$ and $\Delta^2 p \equiv \langle p^2 \rangle - \langle p \rangle^2$). The uncertainty relation (6.2) quantifies the uncertainty principle by giving an absolute bound on the uncertainty product $\Delta q \, \Delta p$.

6.1.1 An abstract Gedanken experiment

What happens when we *attempt* to measure simultaneously the position *and* the momentum of a mechanical system? In 1965 Arthurs and Kelly published a remarkable brief report in the Bell System Technical Journal [8] entitled "On the Simultaneous Measurement of a Pair of Conjugate Observables." They described such an attempt – a Gedanken experiment on a one-dimensional wave packet. Let us understand what is going on there. However, we will not consider Arthurs and Kelly's particular example here. (The reader is referred to Stenholm's excellent article [259] for the details.) We will sketch only the general idea behind this and other schemes for measuring jointly q and p.

Certainly, we cannot measure the position and the momentum simultaneously *and* precisely. Yet literally taken, this fact does not exclude the possibility to observe q and p at the same time with, however, limited accuracy. We could allow for some extra quantum noise to be involved in the measurement process. How can we put this idea into precise terms? Let us assume that we have two meters attached to our system – one for observing q and the other for p, described by the relations

$$\hat{Q}_1 = \hat{q}_s + \hat{A}, \quad \hat{P}_2 = \hat{p}_s + \hat{B}. \tag{6.3}$$

The subscript s refers to the signal, whereas $\hat{Q}_1$ denotes the position read by one meter and $\hat{P}_2$ the momentum read by the other meter. The operators $\hat{A}$ and $\hat{B}$ describe the extra quantum fluctuations necessary for measuring simultaneously the position and the momentum of the signal. We do not need to specify the state of the fluctuations or the particulars of the operators. First, we require only that they carry no preexisting amplitude, that is,

$$\langle \hat{A} \rangle = \langle \hat{B} \rangle = 0. \tag{6.4}$$

Although we cannot measure the intrinsic $\hat{q}_s$ and $\hat{p}_s$ simultaneously, we can read the meters $\hat{Q}_1$ and $\hat{P}_2$ at the same time. We must require that

$$[\hat{Q}_1, \hat{P}_2] = 0. \tag{6.5}$$

As a consequence of the canonical commutator relation

$$[\hat{q}_s, \hat{p}_s] = \mathrm{i} \tag{6.6}$$

for the signal we obtain from Eq. (6.5)

$$-[\hat{A}, \hat{B}] = \mathrm{i} + [\hat{q}_s, \hat{B}] + [\hat{A}, \hat{p}_s]. \tag{6.7}$$

This relation is the only other necessary constraint on the extra fluctuation and shows on a formal level that the introduction of two noncommuting operators $\hat{A}$ and $\hat{B}$ is indeed necessary to satisfy the condition (6.5) for the simultaneous measurement, that is, to compensate the mutual exclusion of $\hat{q}_s$ and $\hat{p}_s$. In classical physics (or in the "classical regime" of a large position-and-momentum scale) we could neglect the extra fluctuations, and we would obtain the simplest possible description of a joint measurement of q and p. So we can agree that the operators $\hat{Q}_1$ and $\hat{P}_2$ defined in Eq. (6.3) describe the outcome of a simultaneous yet imprecise measurement of position and momentum in quantum mechanics.

How precisely do $\hat{Q}_1$ and $\hat{P}_2$ correspond to the actual position $\hat{q}_s$ and momentum $\hat{p}_s$ of the signal? Which limit does Heisenberg's uncertainty relation set? Arthurs and Kelly [8] answered this question in a general and elegant way: They considered the effect of the extra fluctuations on the uncertainty product $\Delta Q_1 \Delta P_2$ of the measured position and momentum values compared with the intrinsic product $\Delta q_s \Delta p_s$. We obtain from definition (6.3) and from Eq. (6.4) that the product $\Delta^2 Q_1 \Delta^2 P_2$ satisfies

$$\Delta^2 Q_1 \Delta^2 P_2 = \Delta^2 q_s \Delta^2 p_s + \langle \hat{A}^2 \rangle \langle \hat{B}^2 \rangle + \Delta^2 q_s \langle \hat{B}^2 \rangle + \Delta^2 p_s \langle \hat{A}^2 \rangle$$

$$\geq \left(\Delta q_s \Delta p_s + \sqrt{\langle \hat{A}^2 \rangle \langle \hat{B}^2 \rangle} \right)^2. \qquad (6.8)$$

In the last line we estimated the arithmetic mean $[\Delta^2 q_s \langle \hat{B}^2 \rangle + \Delta^2 p_s \langle \hat{A}^2 \rangle]/2$ by the geometric mean $\Delta q_s \Delta p_s [\langle \hat{A}^2 \rangle \langle \hat{B}^2 \rangle]^{1/2}$. [We have used the fact that $(a+b)/2 \geq (ab)^{1/2}$ for all real a's and b's.] We apply Heisenberg's uncertainty relation for the intrinsic position q_s and momentum p_s to arrive at

$$\Delta Q_1 \Delta P_2 \geq \frac{1}{2} + \sqrt{\langle \hat{A}^2 \rangle \langle \hat{B}^2 \rangle}. \qquad (6.9)$$

Let us estimate the product of the fluctuations $\langle \hat{A}^2 \rangle$ and $\langle \hat{B}^2 \rangle$. Because their average amplitude vanishes [see Eq. (6.4)], the expectation values $\langle \hat{A}^2 \rangle$ and $\langle \hat{B}^2 \rangle$ describe the variances. We use the general uncertainty relation [74], [236]

$$\langle \hat{A}^2 \rangle \langle \hat{B}^2 \rangle \geq -\frac{1}{4} \langle [\hat{A}, \hat{B}]^2 \rangle \qquad (6.10)$$

to express the fluctuations in terms of the commutator. We assume that the signal state and the fluctuations $\hat{A}$ and $\hat{B}$ are separate, that is, that the corresponding density operators factorize. Consequently, the averages of $[\hat{q}_s, \hat{B}]$ and $[\hat{A}, \hat{p}_s]$ vanish, and we obtain by squaring Eq. (6.7) the expectation value

$$\langle [\hat{A}, \hat{B}]^2 \rangle = -1 + \langle ([\hat{q}_s, \hat{B}] + [\hat{A}, \hat{p}_s])^2 \rangle. \qquad (6.11)$$

Because commutators are anti-Hermitian, the eigenvalues of the operator $[\hat{q}_s, \hat{B}] + [\hat{A}, \hat{p}_s]$ are purely imaginary and the expectation value of $([\hat{q}_s, \hat{B}] + [\hat{A}, \hat{p}_s])^2$ is negative or equal to zero. In this way we find the bound

$$\langle [\hat{A}, \hat{B}]^2 \rangle \leq -1. \qquad (6.12)$$

Finally, we use this estimation in relations (6.9) and (6.10) to arrive at the famous result [8]

$$\Delta Q_1 \Delta P_2 \geq 1. \tag{6.13}$$

This simple relation quantifies the effect of the extra noise involved in a simultaneous yet imprecise measurement of position and momentum. The uncertainty product of the measured Q_1 and P_2 values exceeds the Heisenberg limit (6.2) by a factor of two. As we have seen, this result is rather general and requires few (and quite natural) assumptions.

Given fixed variances $\langle \hat{A}^2 \rangle$ and $\langle \hat{B}^2 \rangle$ of the extra fluctuations in a simultaneous measurement of position and momentum, what are the minimum uncertainty states (with respect to the observed quantities) [166]? Remember that in our analysis we have estimated the arithmetic mean $[\Delta^2 q_s \langle \hat{B}^2 \rangle + \Delta^2 p_s \langle \hat{A}^2 \rangle]/2$ by the corresponding geometric mean, and then we have solely used estimations for the fluctuations $\langle \hat{A}^2 \rangle$ and $\langle \hat{B}^2 \rangle$. Given the latter quantities, the uncertainty product is minimized if the geometric mean equals the arithmetic one, that is, if $\Delta^2 q_s \langle \hat{B}^2 \rangle$ equals $\Delta^2 p_s \langle \hat{A}^2 \rangle$. Also, we must minimize the intrinsic uncertainty product $\Delta q_s \Delta p_s$. According to Pauli's proof (see Section 2.3.) only the squeezed states (2.84) have minimal uncertainty in their intrinsic position and momentum fluctuations. Consequently, the minimum-uncertainty states for the observed joined position-and-momentum values are the squeezed states with

$$\frac{\Delta^2 q_s}{\Delta^2 p_s} = \frac{\langle \hat{A}^2 \rangle}{\langle \hat{B}^2 \rangle} \tag{6.14}$$

or, in terms of the squeezing parameter ζ of Eq. (2.84),

$$\zeta = -\frac{1}{4} \ln(\langle \hat{A}^2 \rangle / \langle \hat{B}^2 \rangle). \tag{6.15}$$

The ratio of the extra fluctuations $\langle \hat{A}^2 \rangle$ and $\langle \hat{B}^2 \rangle$ determines the squeezing for the best adapted state. The less the extra fluctuation of one of the observables $\hat{A}$ or $\hat{B}$ is, the higher is the influence of the intrinsic uncertainty and the higher must be the squeezing of the position or the momentum variance, respectively, for minimizing the uncertainty product.

6.2 Quantum-optical scheme

Quantum optics is the field in which most modern tests of the fundamentals of quantum physics have been performed experimentally. Many classic textbook Gedanken experiments became reality thanks to quantum-optical technology and to the art and the patience of dedicated experimentalists working in this area. How do we bring into being the idea of Arthurs and Kelly? How do we measure simultaneously position and momentum in quantum optics?

Let us first recall what we mean by measurements of position and momentum. We have seen in Section 2.1 that the in-phase and out-of-phase quadrature components $\hat{q}$ and $\hat{p}$ obey the canonical commutation relation (6.1), and, consequently, they share all algebraic properties of mechanical position and momentum operators, respectively. We studied in Section 4.2 how the quadratures can be measured via balanced homodyne detection. Yet in addition, we would like to have a device for making two "copies" of a light beam so that we can measure separately the position quadrature of the first beam and the momentum quadrature of the second "copy." What about using a simple beam splitter? It could split the incident spatial–temporal mode into two parts. We could guide each emerging beam to a homodyne detector, one for measuring $\hat{q}$ on the first beam and the other for measuring $\hat{p}$ on the second field. We must ensure only that the local oscillators of the two homodyne detectors have a phase difference of $\pi/2$. This is readily achieved using a common local oscillator that is split into two parts at a second beam splitter. One partial beam is directed to the first homodyne detector, and the other is phase-shifted via a $\lambda/4$ plate and directed to the second homodyne detector. The scheme appears like two entangled homodyne apparata; see Fig. 6.2. Strictly speaking, four input fields are involved – the signal and a vacuum at the first beam splitter and the local oscillator and a vacuum at the second. Additionally, four output beams are traveling toward the four employed photodetectors. In view of this the apparatus is called an *eight-port homodyne detector*.

This device was used by Walker and Carroll [282], [283] in Cambridge in 1984 to perform the first genuine simultaneous measurement of position and momentum. Although the scheme had precursors in microwave technology [85], these devices had never operated on the quantum level until the pioneering work of Walker and Carroll. Note that apart from the eight-port homodyne detector, other possibilities exist to "copy" a light beam and to measure jointly q and p – the use of a *six-port* [189], [307], for instance. Probably the first who put forward a feasible idea for making "quantum copies" of light were Bandilla and Paul [16], [17] in 1969. They proposed and analyzed theoretically [210] the use of linear *amplification* to measure the quantum-optical phase of a light mode. [The phase problem is considered in Section 6.3.] The amplifier magnifies the signal until it reaches a macroscopic level at which the extra noise involved in a simultaneous measurement of q and p can be safely neglected. However, any linear amplifier is intrinsically noisy [56], so that the amplification noise takes over the role of the extra noise in the general Arthurs and Kelly scheme (6.3). Note also that in *heterodyning* [253], [301] the joint information about q and p is contained in the beating of the signal with a local oscillator of different optical frequency. Interested readers are referred to review article [163], where

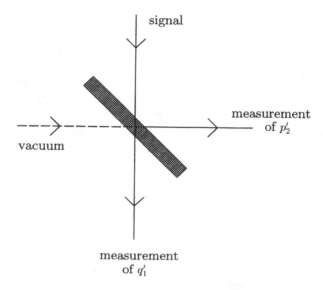

Fig. 6.1. Simultaneous measurement of position and momentum quadratures. The incident signal is split into two emerging beams. Each beam represents an independent system. In this case quantum mechanics does not raise any objections to measurements of the position quadrature on one beam and of the momentum on the other. However, the vacuum field entering the apparatus via the "unused" second port of the beam splitter introduces extra noise. The uncertainty principle is not violated but taken literally, that is, we can measure the position and the momentum simultaneously but not precisely.

amplification and heterodyning are compared with the beam-splitting idea in some detail to show their common roots. Here we will restrict our attention to the eight-port homodyne detector only. This device is probably the clearest and most elegant scheme to measure simultaneously position and momentum in quantum optics.

6.2.1 Heisenberg picture

Why is quantum mechanics not violated in the joint measurement of canonically conjugate quadratures? How is this scheme related to the general idea of Arthurs and Kelly [8]? What does the eight-port homodyne detector actually measure? To answer all these questions we need to understand only the action of the first beam splitter, where the signal is divided into two parts. The rest of the device serves to perform only the homodyne measurements on the two emerging beams (provided, of course, that the local oscillator is strong).

Roughly speaking, beam splitting is always "noisy." The incident photons are distributed as whole energy "lumps" to the two emerging beams. Yet the average

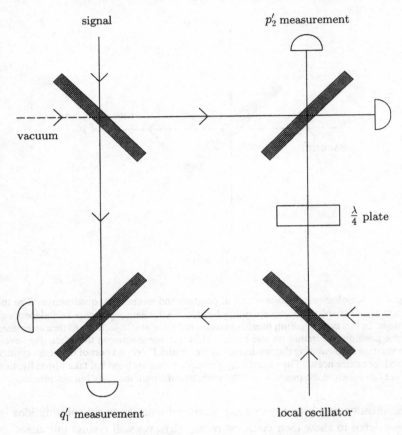

signal p_2' measurement

vacuum

$\frac{\lambda}{4}$ plate

q_1' measurement   local oscillator

Fig. 6.2. Eight-port homodyne detector.

intensity ratio of the transmitted and the reflected beam is a constant given by the ratio of the transmittance τ and the reflectance ϱ of the beam splitter. To resolve this conflict between the wavelike distribution of the average intensities and the appearance of discrete particles, the signal photons are distributed as single units but at random. Each photon is transmitted with the probability τ^2 and reflected with the probability ϱ^2. This randomness causes some additional noise in the simultaneously measured quadratures. Alternatively, we may understand the extra detection noise as being caused by the vacuum entering the second (the "unused") port of the beam splitter. These vacuum fluctuations contaminate the signal field, so that the jointly measured quadratures are only fuzzy pictures of the intrinsic position and momentum quantities. In this way the violation of Heisenberg's uncertainty principle is avoided.

To put these words into precise terms, let us apply the simple quantum theory of beam splitting developed in Section 4.1. Beam splitting is based on optical interference, that is, on the superposition of the incident fields. In the Heisenberg picture the annihilation operators $\hat{a}_1'$ and $\hat{a}_2'$ of the emerging beams are linear transformations of the operators $\hat{a}_1$ and $\hat{a}_2$ for the incident modes

$$\begin{pmatrix} \hat{a}_1' \\ \hat{a}_2' \end{pmatrix} = \begin{pmatrix} \tau & -\varrho \\ \varrho & \tau \end{pmatrix} \begin{pmatrix} \hat{a}_1 \\ \hat{a}_2 \end{pmatrix}. \tag{6.16}$$

For simplicity we have assumed that the beam-splitting matrix is real. We note that this situation can always be achieved by redefining the reference phases of the mode operators $\hat{a}_1$, $\hat{a}_2$ and $\hat{a}_1'$, $\hat{a}_2'$. The relation

$$\tau^2 + \varrho^2 = 1 \tag{6.17}$$

between the transmittance τ and the reflectance ϱ accounts for the energy conservation of a lossless beam splitter or, equivalently, for the fact that photons are either transmitted (with the probability τ^2) or reflected (with the probability ϱ^2). The eight-port homodyne detector measures the q quadrature of the first emerging beam and the p quadrature of the second. Because $\hat{a}$ equals $2^{-1/2}(\hat{q} + i\hat{p})$, these quantities are given by the simple expressions*

$$\hat{q}_1' = \tau\hat{q}_1 - \varrho\hat{q}_2, \qquad \hat{p}_2' = \varrho\hat{p}_1 + \tau\hat{p}_2. \tag{6.18}$$

As a consequence of the canonical commutation relation (6.1) for the operator pairs $\hat{q}_1$, $\hat{p}_1$ and $\hat{q}_2$, $\hat{p}_2$, the measured quadratures $\hat{q}_1'$ and $\hat{p}_2'$ do commute. This fact is not surprising, for otherwise we simply could not measure the two quantities simultaneously. Formula (6.18) shows clearly how the eight-port homodyne detector brings into being the central idea (6.3) of Arthurs and Kelly. The measured q quadrature is proportional to the intrinsic position $\hat{q}_1$, except for the noise term $-\varrho\hat{q}_2$. Because only a fraction of τ^2 of the incident intensity reaches the q detector, the measured position quadrature is reduced by the factor of τ. The extra quantum noise originates from the field entering the second port of the beam splitter. The noise is enhanced for a low transmittance τ and reduced for a highly transmitting beam splitter. Similarly, the p detector measures $\varrho\hat{p}_1$, apart from the noise contribution $\tau\hat{p}_2$.

So the quantum-optical version of Arthurs and Kelly's Gedanken experiment is as simple as this: Split a beam into two parts and measure simultaneously the position quadrature on one beam and the momentum quadrature on the other.

---

*Note [97] that $\hat{q}_1'$ and $\hat{p}_2'$ are exactly the observables in the original Einstein–Podolsky–Rosen Gedanken experiment [36], [84].

The measured quantities are proportional to the operators

$$\hat{Q}_1 = \hat{q}_1 - \frac{\varrho}{\tau}\hat{q}_2, \quad \hat{P}_2 = \hat{p}_1 + \frac{\tau}{\varrho}\hat{p}_2, \tag{6.19}$$

that is, to the Arthurs and Kelly variables (6.3). The mode entering the second (the "unused") port of the beam splitter brings about just the right quantity of extra quantum fluctuations required for not violating Heisenberg's uncertainty principle. We have seen in Section 6.1.1 that one effect of this extra noise is the doubling (6.13) of the uncertainty product.

6.2.2 Phase-space density and squeezing

What is the probability distribution $\mathrm{pr}(Q_1, P_2)$ for the simultaneously measured position-and-momentum values? From this distribution we could gain much more detailed information about the effect of the extra quantum noise involved than quantified in the uncertainty product only. Again, we use our simple quantum theory of beam splitting to calculate $\mathrm{pr}(Q_1, P_2)$.

Let us describe the state of the incident signal by the density operator $\hat{\rho}$ and the associated Wigner function $W(q_1, p_1)$. Additionally, we must take into account the light beam entering the second input port of the beam splitter. In most experiments this beam would just "not exist" classically, meaning in quantum optics that the second incident mode is a vacuum with the Wigner function (3.32)

$$W_0(q_2, p_2) = \frac{1}{\pi}\exp\left(-q_2^2 - p_2^2\right). \tag{6.20}$$

According to Eq. (4.34) the beam splitter transforms the total Wigner function $W(q_1, p_1, q_2, p_2)$ of the two incident beams as if $W(q_1, p_1, q_2, p_2)$ were a classical probability distribution for the quadratures q_1, p_1 and q_2, p_2. This is in formulas

$$W'(q_1, p_1, q_2, p_2) = W(q_1', p_1')W_0(q_2', p_2') \tag{6.21}$$

with the changed variables [inversely to Eq. (6.16)]

$$\begin{pmatrix} q_1' \\ q_2' \end{pmatrix} = \begin{pmatrix} \tau & \varrho \\ -\varrho & \tau \end{pmatrix}\begin{pmatrix} q_1 \\ q_2 \end{pmatrix}, \quad \begin{pmatrix} p_1' \\ p_2' \end{pmatrix} = \begin{pmatrix} \tau & \varrho \\ -\varrho & \tau \end{pmatrix}\begin{pmatrix} p_1 \\ p_2 \end{pmatrix}. \tag{6.22}$$

As a fundamental property (3.1) of the Wigner function, the probability distribution $\mathrm{pr}(q_1, p_2)$ is given by integrating $W'(q_1, p_1, q_2, p_2)$ with respect to the unobserved quantities p_1 and q_2, that is, by

$$\mathrm{pr}(q_1, p_2) = \int_{-\infty}^{+\infty}\int_{-\infty}^{+\infty} W'(q_1, p_1, q_2, p_2)\,dp_1\,dq_2. \tag{6.23}$$

To find a tale-telling expression for $\mathrm{pr}(q_1, p_2)$ we simply change the variables in the integration (6.23). We use

$$q = q_1' = \tau q_1 + \varrho q_2, \qquad p = p_1' = \tau p_1 + \varrho p_2 \qquad (6.24)$$

instead of p_1 and q_2. Obviously,

$$q_2' = -\varrho q_1 + \tau q_2 = \frac{1}{\varrho}(-q_1 + \tau q), \qquad (6.25)$$

where we have used energy conservation (6.17). In a similar way we get

$$p_2' = \frac{1}{\tau}(p_2 - \varrho p). \qquad (6.26)$$

Splitting a beam into two parts means distributing the incident intensity to the two emerging beams (with a ratio of τ/ϱ). To compensate for this intensity loss we rescale the position-and-momentum variables

$$Q_1 = \frac{q_1}{\tau}, \qquad P_2 = \frac{p_2}{\varrho} \qquad (6.27)$$

and their probability distribution

$$\mathrm{pr}(Q_1, P_2) = (\varrho\tau)^{-1}\mathrm{pr}(q_1, p_2). \qquad (6.28)$$

The prefactor $(\varrho\tau)^{-1}$ appears because $\mathrm{pr}(Q_1, P_2)$ is a probability density. The distribution $\mathrm{pr}(Q_1, P_2)$ of the scaled variables Q_1 and P_2 describes the information we infer about the "intrinsic" phase-space density.

Taking all these calculations and definitions (6.20–6.28) into account, we obtain the simple result

$$\mathrm{pr}(Q_1, P_2) = \int_{-\infty}^{+\infty} \int_{-\infty}^{+\infty} W(q, p) \qquad (6.29)$$

$$\times \frac{1}{\pi} \exp\left[-\frac{\tau^2}{\varrho^2}(q - Q_1)^2 - \frac{\varrho^2}{\tau^2}(p - P_2)^2\right] dq\, dp.$$

The probability distribution $\mathrm{pr}(Q_1, P_2)$ for the simultaneously measured position-and-momentum quantities is a filtered Wigner function. It resolves $W(q, p)$ within an ellipse in phase space; see Fig. 6.3. Let us consider a balanced beam splitter with $\tau = \varrho$. We look at the definition (3.46) of the Q function and notice immediately that the measured phase-space density is exactly this quasiprobability distribution. *The eight-port homodyne detector measures the Q function!* Note that this feature is shared by a remarkable number of other schemes for measuring simultaneously position and momentum [152], [244], [253], [307] and in particular by the original example of Arthurs and Kelly [8], [259] (for balanced meters).

In the unbalanced case the Q function is "squeezed". This "squeezed Q function" is called a *Husimi function* [120]. To quantify the measurement-induced

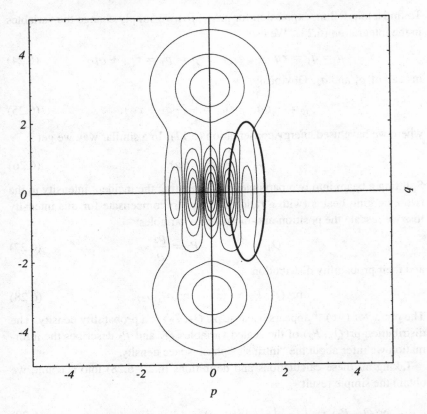

Fig. 6.3. The simultaneously measured phase-space density resolves the Wigner function within an ellipse in phase space.

squeezing, we take advantage of the overlap formula (3.20) and the Wigner representation (3.34), (3.39) of the squeezing (2.82) and the displacement operator (2.48), respectively. We see that the measured phase-space density is a quantum overlap [155], [283]

$$\text{pr}(Q_1, P_2) = \frac{1}{2\pi}\text{tr}\{\hat{\rho}|\alpha, \zeta\rangle\langle\alpha, \zeta|\} \tag{6.30}$$

with the squeezed state

$$|\alpha, \zeta\rangle = \hat{D}(\alpha)\hat{S}(\zeta)|0\rangle \tag{6.31}$$

characterized by the displacement

$$\alpha = 2^{-1/2}(Q_1 + iP_2) \tag{6.32}$$

and the squeezing parameter

$$\zeta = \ln(\tau/\varrho). \tag{6.33}$$

To explain this squeezing effect we note that for a large transmittance τ the precision of the Q_1 quadrature is enhanced, whereas the precision of P_2 is reduced. The reason is simply that in this case the intrinsic position $\hat{q}_1$ in $\hat{Q}_1$ is less contaminated by the vacuum fluctuations entering the "unused" port of the beam splitter, whereas p_1 suffers accordingly. And of course, for a small transmittance τ the situation is reversed. To be more quantitative, we calculate the measured position distribution $\mathrm{pr}(Q_1)$ by integrating $\mathrm{pr}(Q_1, P_2)$ with respect to P_2. We obtain immediately from Eq. (6.29)

$$\mathrm{pr}(Q_1) = \pi^{-1/2} |\tau/\varrho| \int_{-\infty}^{+\infty} \mathrm{pr}(q) \exp[-(\tau/\varrho)^2 (q - Q_1)^2] \, dq, \tag{6.34}$$

because the marginals of the Wigner function are the quadrature distributions (3.1). The squeezed position fluctuations of the incident vacuum filter the distribution $\mathrm{pr}(q)$ and limit the resolution of the intrinsic position quadrature. Evidently, for the momentum distribution we obtain the similar result

$$\mathrm{pr}(P_2) = \pi^{-1/2} |\varrho/\tau| \int_{-\infty}^{+\infty} \mathrm{pr}(p) \exp[-(\varrho/\tau)^2 (p - P_2)^2] \, dp. \tag{6.35}$$

If the precision of the measured Q_1 quadrature is high, the precision of P_2 is accordingly poor. In an extreme case the beam splitter is completely transmitting, so that the Q_1 homodyne detector receives the full signal. The Q_1 resolution is perfect, but, on the other hand, the P_2 detector is detached from the signal and, consequently, measures no signal information anymore. By changing the transmittance we could squeeze the resolution of the signal Wigner function, but we can never make the total effect of the extra quantum fluctuations arbitrarily small.

Yes, we can see experimentally an overall phase-space picture of a quantum system, but the picture is fuzzy. This is the price to be paid if we dare to measure simultaneously the position and the momentum, that is, all phase-space properties, in a single experiment. However, if we prefer to observe the phase-space aspects one at a time (in the form of phase-shifted quadratures) we circumvent this problem. In this case we can indeed tomographically infer the Wigner function, as we have seen in Chapter 5.

6.2.3 Optical interference and squared Wigner function

So far we have assumed that we split the incident signal to measure the position quadrature on one of the emerging beams and the momentum quadrature on the other. Although "unused," the second input port of the beam splitter allows the vacuum fluctuations to sneak in. This picture has helped us to understand why the apparatus does not violate Heisenberg's uncertainty principle. What happens if we let the signal interfere with a second incident beam instead of just splitting it? This second field might be a squeezed vacuum or a coherent beam of laser light, for instance. To describe the effect of the interference with "something" instead of "nothing" we can, fortunately, perform the same calculations as in the previous case. The only difference is that we do not specify the state $\hat{\rho}_R$ of the second incident field and the associated Wigner function $W_R(q_2, p_2)$. To calculate the measured phase-space density we simply repeat the procedure of Section 6.2.2 and obtain the result

$$\mathrm{pr}(Q_1, P_2) = \int_{-\infty}^{+\infty} \int_{-\infty}^{+\infty} W(q, p) W_{SQR}(q - Q_1, p - P_2) \, dq \, dp \quad (6.36)$$

with

$$W_{SQR}(q, p) = W_{QR}\left(\frac{\tau}{\varrho}q, \frac{\varrho}{\tau}p\right) \quad (6.37)$$

and

$$W_{QR}(q, p) = W_R(q, -p) \quad (6.38)$$

So quite generally, the probability distribution $\mathrm{pr}(Q_1, P_2)$ for the simultaneously measured position-and-momentum values is a filtered Wigner function. Following Popper [222], Wódkiewicz [293], [294] called the expression (6.36) a *propensity*. Two operations relate the filter function $W_{SQR}(q, p)$ to the Wigner function $W_R(q, p)$ of the second incident beam. The first one is the already familiar squeezing (6.37) brought about by using an unbalanced beam splitter. Additionally, the momenta p are inverted in Eq. (6.38). We have not noticed this feature in the case of beam splitting because the vacuum Wigner function is inversion invariant. In classical optics the inversion of the p quadrature components means a *phase conjugation* because p is proportional to the imaginary part of the complex wave amplitude α. The signal interferes optically with the second beam, and so the phase-conjugated wave amplitude of the latter enters the interference pattern. We would expect that $W_{QR}(q, p)$ is the Wigner function for the complex conjugate of the density matrix

$$\hat{\rho}_{QR} \equiv \hat{\rho}_R^* \quad (6.39)$$

in position representation. [We mean by this definition that the density matrix $\langle q|\hat{\rho}_{QR}|q\rangle$ should be the complex conjugate of $\langle q|\hat{\rho}_R|q\rangle$.] That Eq. (6.39) is indeed correct is easily seen using Wigner's formula (3.17)

$$W_{QR}(q, p) = \frac{1}{2\pi} \int_{-\infty}^{+\infty} \exp(ipx) \left\langle q - \frac{x}{2} \middle| \hat{\rho}_R \middle| q + \frac{x}{2} \right\rangle^* dx$$

$$= \frac{1}{2\pi} \int_{-\infty}^{+\infty} \exp(ipx) \left\langle q + \frac{x}{2} \middle| \hat{\rho}_R \middle| q - \frac{x}{2} \right\rangle dx$$

$$= W_R(q, -p). \tag{6.40}$$

In the last step we have replaced x by $-x$ in the integration.

Aharonov, Albert, and Au [6] called the complex conjugate $\hat{\rho}_{QR}$ of $\hat{\rho}_R$ the density matrix of the *quantum ruler*. The ruler, essentially the second incident field, probes the Wigner function of the signal. The filter function in $\mathrm{pr}(Q_1, P_2)$ is the squeezed and displaced Wigner function $W_{QR}(q_2, p_2)$ of the quantum ruler. When the second input port of the beam splitter is "unused," the ruler is in the vacuum state and we obtain Eq. (6.29). As in this special case, we could squeeze the resolution of the signal Wigner function by changing the transmittance of the beam splitter. However, as we have seen in the discussion of Eq. (3.28), any physically meaningful Wigner function cannot be highly peaked and in particular cannot approach a two-dimensional delta function. Of course, this applies also to the Wigner function of the quantum ruler. So the resolution of the filtering (6.36) is always limited, and we can never measure the true Wigner function directly as a probability distribution [apart from the mere fact that $W(q, p)$ might be negative.]

Beam splitters are noisy "copy machines" for quantum light fields, and therefore they cause extra quantum fluctuations in a simultaneous measurement of position-and-momentum quadratures. What would happen, however, if we had already two perfect copies of a light beam and let them interfere [161]? Imagine that the first incident beam is in a pure state described by Schrödinger's position wave function $\psi(q_1)$, whereas the second beam is just in the complex conjugate state $\psi^*(q_2)$. Both light fields should interfere at a balanced beam splitter with $\tau = \varrho = 2^{-1/2}$, and we measure q on the first and p on the second beam. According to our quantum theory of beam splitting [155] (see Section 4.1.2), the total wave function $\psi'(q_1, q_2)$ of the emerging fields is the rotated wave function of the incident beams, that is,

$$\psi'(q_1, q_2) = \psi[2^{-1/2}(q_1 - q_2)] \psi^*[2^{-1/2}(q_1 + q_2)]. \tag{6.41}$$

Because the momentum is measured on the second beam, it is advantageous to express ψ' in the momentum representation with respect to p_2

$$\tilde{\psi}'(q_1, p_2) = \frac{1}{\sqrt{2\pi}} \int_{-\infty}^{+\infty} \psi'(q_1, q_2) \exp(-ip_2 q_2) \, dq_2. \tag{6.42}$$

According to Born's interpretation of the wave function, the modulus squared of $\tilde{\psi}'(q_1, p_2)$ gives the probability distribution $\mathrm{pr}(q_1, p_2)$ of the measured q_1 and p_2 values, that is,

$$\mathrm{pr}(q_1, p_2) = \frac{1}{2\pi} \left| \int_{-\infty}^{+\infty} \psi[2^{-1/2}(q_1 - q_2)]\psi^*[2^{-1/2}(q_1 + q_2)] \right.$$

$$\left. \times \exp(-\mathrm{i}p_2 q_2)\, dq_2 \right|^2. \tag{6.43}$$

We introduce the scaled variables

$$x = 2^{-1/2}q_2, \quad Q_1 = 2^{-1/2}q_1, \quad P_2 = 2^{-1/2}p_2 \tag{6.44}$$

and, according to Eq. (6.28), the corresponding probability distribution $\mathrm{pr}(Q_1, P_2)$ to obtain

$$\mathrm{pr}(Q_1, P_2) = \frac{1}{\pi} \left| \int_{-\infty}^{+\infty} \psi(Q_1 - x)\psi^*(Q_1 + x) \exp(-\mathrm{i}P_2 x)\, dx \right|^2. \tag{6.45}$$

We glance at Wigner's formula (3.17) and realize immediately that $\mathrm{pr}(Q_1, P_2)$ is essentially the modulus squared of the Wigner function. Because the Wigner function is real, the modulus squared is just the square, and we get the result [161]

$$\mathrm{pr}(Q_1, P_2) = 2\pi W^2(Q_1, -P_2). \tag{6.46}$$

Although we cannot measure the Wigner function directly as a probability distribution, we can, in principle, measure the square of $W(q, p)$! However, for this measurement we need to have already two copies of the light beam, one in the state $\psi(q_1)$ and the other in the conjugate state $\psi^*(q_2)$.

Can we copy a quantum state? No! *Copying* (also called *cloning*) violates the superposition principle [76], [296]. To see this in the most elementary way, imagine a "quantum copy machine" described by some unitary transformation $\hat{X}$. Let $\hat{X}$ act on the wave function $\psi(q)$ and on some auxiliary system $\psi_{AUX}(q_{AUX})$ that serves as a tabula rasa for making copies. The quantum copy machine should produce $\psi(q_1)$ and $\psi^*(q_2)$, that is,

$$\hat{X}\psi(q)\psi_{AUX}(q_{AUX}) = \psi(q_1)\psi^*(q_2), \tag{6.47}$$

for all quantum states $\psi(q)$ and their superpositions. Imagine we represent $\psi(q)$ as

$$\psi(q) = c_1\psi_1(q) + c_2\psi_2(q) \tag{6.48}$$

with some wave functions $\psi_1(q)$ and $\psi_2(q)$. According to the superposition principle the quantum copy machine would produce the state

$$\hat{X}\psi(q)\psi_{AUX}(q_{AUX}) = c_1\hat{X}\psi_1(q)\psi_{AUX}(q_{AUX})$$

$$+ c_2\hat{X}\psi_2(q)\psi_{AUX}(q_{AUX})$$

$$= c_1\psi_1(q_1)\psi_1^*(q_2) + c_2\psi_2(q_1)\psi_2^*(q_2)$$

$$\neq \psi(q_1)\psi^*(q_2). \tag{6.49}$$

This is not the quantum copy of $\psi(q)$! Cloning is impossible! (For a more refined version of the *no-cloning theorem* see Ref. [23].) So if we had two copies of a quantum system, we could measure the squared Wigner function as a joint probability distribution for position and momentum. However, we can never make these copies from a single system. Our analysis shows yet another aspect of the problems involved in measuring simultaneously the position and the momentum.

6.3 Quantum-optical phase

In quantum optics we encounter another famous pair of canonically conjugate variables – the *photon number* and the *optical phase*. Indeed, they describe most clearly the distinction between the particle and the wave aspects of light. The canonical pair of number and phase was at the very heart of the first version of quantum electrodynamics developed by Dirac himself [77]. And yet, as an irony of history, the concept of quantum-optical phase suffered from deeply rooted theoretical difficulties for a long time after the final establishment of QED. The debate about the phase problem created a huge literature of interesting and/or odd papers; see Lynch's excellent Critical Review [186]. What is the problem? Let us assume that the photon number $\hat{n}$ and the Hermitian phase operator $\hat{\phi}$ are canonical conjugates in the same way as the position $\hat{q}$ and the momentum $\hat{p}$ are, that is, we require*

$$[\hat{n}, \hat{\phi}] = i.$$

We define the operator

$$\hat{E}(\nu) \equiv \exp(-i\nu\hat{\phi})$$

for real ν's and see easily that

$$\frac{d}{d\nu}\hat{E}^{-1}(\nu)\hat{n}\hat{E}(\nu) = \hat{E}^{-1}(\nu)[i\hat{\phi}, \hat{n}]\hat{E}(\nu) = 1.$$

*To avoid confusion we omit the equation numbers for questionable expressions.

Consequently, the operator $\hat{E}(\nu)$ shifts the photon number

$$\hat{E}^{-1}(\nu)\hat{n}\hat{E}(\nu) = \hat{n} + \nu.$$

Suppose that $|n\rangle$ is an eigenstate of the photon number (a Fock state). Because of

$$\hat{n}\hat{E}(\nu)|n\rangle = \hat{E}(\nu)(\hat{n} + \nu)|n\rangle = (n + \nu)\hat{E}(\nu)|n\rangle$$

the state $\hat{E}(\nu)|n\rangle$ would be an eigenstate of $\hat{n}$ as well. The operator $\hat{E}(\nu)$ shifts the photon-number states, too. However, the eigenvalue $n + \nu$ is not necessarily an integer, and so the spectrum of the photon-number operator becomes continuous and unbounded. This is impossible! Louisell [180] noticed another simple contradiction in the canonical commutation relation between number and phase. We write $[\hat{n}, \hat{\phi}] = \mathrm{i}$ in the number-state basis

$$\langle n|(\hat{n}\hat{\phi} - \hat{\phi}\hat{n}) \,|\, n'\rangle = \mathrm{i}\langle n \,|\, n'\rangle = \mathrm{i}\delta_{nn'}.$$

On the other hand

$$\langle n|(\hat{n}\hat{\phi} - \hat{\phi}\hat{n})|n'\rangle = (n - n')\langle n|\hat{\phi}|n'\rangle,$$

which gives

$$0 = \mathrm{i}$$

for $n = n'$. Of course, we could apply the same reasoning to the position $\hat{q}$ and the momentum $\hat{p}$, but, fortunately, the limit $q \to q'$ in the scalar product $\langle q \,|\, q'\rangle$ does not exist. Because photon numbers are discrete integers, we do not have this excuse anymore.

Clearly, something was wrong in our "canonical quantization" of the optical phase. Note that there are many more and deeper theoretical problems [186] with the concept of phase than come to the surface in our naive approach. Only recently, Pegg and Barnett [19], [22], [213], [214] succeeded in constructing Hermitian phase operators in a sequence of finite-dimensional Hilbert spaces. Vaccaro [273] changed the structure of the Hilbert space to embed a Hermitian phase operator. Additionally, the Fock space has been extended to negative "photon numbers" [18], [196] for the sole purpose of constructing phase operators. However, all these concepts coincide in their predicted expectation values for physically relevant states. These concepts also agree with another approach [111], [254] to quantum-optical phase based on phase distributions instead of phase operators. We will motivate and develop this concept in Section 6.3.1.

But is the quantum-optical phase really an observable quantity? Can we measure phase? Certainly, every interferometer measures an optical interference pattern that depends on the phase of the signal (with respect to a certain reference). In classical optics, the phase φ is just the argument of the complex wave amplitude α such that

$$\alpha = |\alpha| \exp(i\varphi) \tag{6.50}$$

or, expressed in terms of the quadratures,

$$\cos \varphi = \frac{q}{\sqrt{q^2 + p^2}}, \qquad \sin \varphi = \frac{p}{\sqrt{q^2 + p^2}}. \tag{6.51}$$

However, in the quantum regime both the phase "φ" and the amplitude "$|\alpha|$" fluctuate, and, unfortunately, we cannot clearly separate the phase and the intensity noise. Only for a large amplitude does the intensity noise become negligible compared with the intensity itself, and we can measure the phase precisely. But then we are leaving the "quantum sector" in which we are interested. To put the problem into the language of this chapter, a typical phase measurement is already a simultaneous measurement of phase and photon number. From our intuition about the quantum nature of the optical phase we would expect that number and phase are complementary. Consequently, realistic phase measurements cannot measure the quantum-optical phase with perfect accuracy.

Why not turn the tables and define the quantum-optical phase operationally? The way we measure phase defines what we mean by it! Bandilla and Paul [16], [17] were the first who put forward this "practically minded" approach to the phase problem (based on linear amplification), and Paul analyzed the theory in his comprehensive paper [210]. First experiments [104], [105] were performed in the early seventies. Shapiro and Wagner [253] proposed another type of phase measurements based on heterodyne detection [163]. Finally, Noh, Fougères, and Mandel [197]–[201] applied a balanced eight-port homodyne detector (see Section 6.2) for defining phase operationally. In fact, in classical optics this device measures the phase difference between the incident signal and the reference beam, the local oscillator. Classically, the apparatus determines the quadratures q and p (with respect to the local-oscillator phase) or, in other words, the complex wave amplitude α. Then the phase is simply given by the relation (6.50) or (6.51). Noh, Fougères, and Mandel were interested in the quantum regime, where they measured the quantity corresponding to the classical optical phase and considered it the operationally defined quantum-optical phase. This remarkable experiment initiated a series of theoretical studies (see again the critical review [186]) and created a number of interesting results and ideas in the course of the debate. At first glance, however, the operational approach to quantum phase introduces an element of subjectivity into the definition of a physical quantity, especially for weak fields in the quantum regime. So it was surprising [97], [152], [244] that most of the operational definitions yield identical results under reasonable, simplifying assumptions.

Let us analyze the Noh–Fougères–Mandel experiment. We assume that the local oscillator should provide a perfect reference for the phase measurement. For this the LO must be strong compared to the signal, so strong that we can treat the local oscillator classically. In this case the balanced eight-port homodyne detector measures the Q function as the probability distribution for q_1 and p_2. See Section 6.2.2. The quadratures q_1 and p_2 correspond to the real and imaginary part of the complex amplitude α, and the phase φ is the argument of α. All statistical properties of φ are described by the phase distribution $\text{pr}(\varphi)$. To calculate $\text{pr}(\varphi)$ we simply discard the amplitude information in the measured Q function by introducing polar coordinates, integrating with respect to the radius $r = |\alpha|$, and obtaining

$$\text{pr}(\varphi) = \int_0^\infty Q(r \cos \varphi, r \sin \varphi) r \, dr. \qquad (6.52)$$

Note that the other mentioned phase-measurement schemes [16], [17], [197], [253] yield the same phase distribution (also called the Q phase in the jargon of quantum optics). Hence they are considered physically equivalent. How is the measured (the operationally defined) phase related to the canonical (the intrinsic) phase?

6.3.1 Canonical phase distribution

What is the canonical phase? To find a satisfying answer we consider phase distributions instead of phase operators. Because a probability distribution of phase contains all statistical information on the phase properties for a given state, this distribution represents the physical quantity phase completely. Realistic measurements of phase [154] always involve extra noise beyond that due to the intrinsic quantum phase fluctuations described by the canonical phase distribution. Consequently, we need a general method that allows the description of phase in the presence of noise. For this we use *probability operator measures* (POMs) [46], [111], [254]. We introduce the probability distribution $\text{pr}(\varphi)$, $\varphi \in (-\pi, +\pi]$, as

$$\text{pr}(\varphi) = \text{tr}\{\hat{\rho}\hat{\Pi}(\varphi)\}, \qquad (6.53)$$

where $\hat{\rho}$ is the density matrix and $\hat{\Pi}(\varphi)$ denotes a set of suitable operators parameterized by the phase variable φ. Because probability distributions are real functions, $\hat{\Pi}(\varphi)$ must be Hermitian

$$\hat{\Pi}^\dagger(\varphi) = \hat{\Pi}(\varphi). \qquad (6.54)$$

A probability distribution is normalized to unity. Consequently, the set of operators $\hat{\Pi}(\varphi)$ must be normalized as well

$$\int_{-\pi}^{+\pi} \hat{\Pi}(\varphi)\, d\varphi = 1. \tag{6.55}$$

Lastly, a probability distribution is nonnegative,

$$\mathrm{pr}(\varphi) \geq 0, \tag{6.56}$$

which implies that the eigenvalues of $\hat{\Pi}(\varphi)$ must be nonnegative, too. This property together with (6.54) and (6.55) is sufficient to identify $\hat{\Pi}(\varphi)$ as a density operator, which is often laxly called a *phase state*. One important point, however, is that the operator $\hat{\Pi}(\varphi)$ might be unnormalizable, although we note that such states can be represented in larger spaces, for instance in the rigged Hilbert space [33]. Another important point is that a phase state $\hat{\Pi}(\varphi)$ can be a mixed state, which simply means that the measure of φ is not precise, and so the probability distribution $\mathrm{pr}(\varphi)$ represents the result of a noisy measurement of φ. For this reason, we call $\mathrm{pr}(\varphi)$ a *noisy phase distribution* and $\hat{\Pi}(\varphi)$ a *mixed phase state*.

So far, we have briefly summarized the general properties of POMs. Now, we turn to two specific requirements for a noisy quantum-phase distribution. Our goal is the definition of quantum-optical phase as the canonically conjugate variable with respect to photon number. We wish to treat number and phase similarly to the basic canonically conjugate variables position and momentum. As we have seen in the previous section, we must not extend the canonical commutation relation (6.1) for position and momentum operators $\hat{q}$ and $\hat{p}$ to a relation for number $\hat{n}$ and phase $\hat{\varphi}$. Here we are considering not phase operators but phase distributions. In defining phase as canonically conjugate to photon number, we should translate some typical properties of position and momentum distributions into the language of number and phase and regard them as being fundamental. What do we mean by canonically conjugate? Position and momentum properties are mutually exclusive. Therefore, they must be independent. If we change the position, the momentum properties are not affected, and, of course, vice versa. We may regard this mutual independence of the canonically conjugate variables as being fundamental and require the same for number and phase.

We require that the phase-distribution function of a single mode satisfy the following axioms: (A) *A phase shifter shifts the phase distribution* [215]. (B) *A number shifter does not change the phase distribution.* (complementarity) [162]. To be explicit, a phase shifter is represented by the unitary transformation operator (2.6)

$$\hat{U}(\phi) = \exp\left(-i\phi\hat{a}^\dagger\hat{a}\right). \qquad (6.57)$$

Axiom (A) thus means that

$$\mathrm{pr}'(\varphi) \equiv \mathrm{tr}\left\{\hat{U}(\phi)\hat{\rho}\hat{U}^\dagger(\phi)\hat{\Pi}(\varphi)\right\}$$

$$= \mathrm{pr}(\varphi + \phi)$$

$$= \mathrm{tr}\{\hat{\rho}\hat{\Pi}(\varphi + \phi)\}. \qquad (6.58)$$

A number shifter is expressed by the operator

$$\hat{E} \equiv \sum_{n=0}^{\infty} |n+1\rangle\langle n|. \qquad (6.59)$$

It shifts the photon-number distribution up by one step. (The operator $\hat{E}$ is called the *Susskind–Glogower exponential phase operator* $\widehat{\exp}(-i\varphi)$.) Axiom (B) thus requires

$$\mathrm{pr}'(\varphi) \equiv \mathrm{tr}\left\{\hat{E}\hat{\rho}\hat{E}^\dagger\hat{\Pi}(\varphi)\right\} = \mathrm{pr}(\varphi). \qquad (6.60)$$

Both axioms (A) and (B) together determine a phase distribution. What do they mean physically? Axiom (A) is almost trivial. We require only that the phase distribution indeed reflect the basic features of quantum phase, that is, that a phase shifter is a phase-distribution shifter. Naturally, many phase-sensitive quantities have the property (A). Axiom (B) is more specific. It means that the distribution function $\mathrm{pr}(\varphi)$ contains the properties of quantum phase and nothing else. It must not reflect any properties of the canonically conjugate variable, the photon number. Hence (B) means that phase should be complementary to photon number. We also note, however, that if a particular distribution function $\mathrm{pr}(\varphi)$ satisfies the axioms, then so does the weighted average $p_1 \mathrm{pr}(\varphi) + p_2 \mathrm{pr}(\varphi + \delta)$ of this function and the phase-shifted distribution $\mathrm{pr}(\varphi + \delta)$, which describes uncertainty in the reference phase. We interpret this as the axioms allow for a noisy measure of phase. The nature of this noise is special in that the resulting distribution still satisfies the axioms of complementarity. Thus, our approach here contains, in essence, the basic prescription for describing a noisy measurement of phase without contamination from the complementary, observable, photon number.

Now we consider the detailed consequences of both axioms. We express the noisy phase probability distribution in the Fock basis

$$\mathrm{pr}(\varphi) = \sum_{n,m=0}^{\infty} \langle m|\hat{\Pi}(\varphi)|n\rangle\langle n|\hat{\rho}|m\rangle. \qquad (6.61)$$

We use axiom (A)

$$\text{pr}(\varphi) = \text{pr}[0 + \varphi]$$

$$= \sum_{n,m=0}^{\infty} \langle m|\hat{\Pi}(0)|n\rangle \langle n|\hat{U}(\varphi)\hat{\rho}\hat{U}^{\dagger}(\varphi)|m\rangle$$

$$= \sum_{n,m=0}^{\infty} \langle m|\hat{\Pi}(0)|n\rangle \exp[\mathrm{i}(m-n)\varphi]\langle n|\hat{\rho}|m\rangle, \qquad (6.62)$$

define the coefficients

$$B_{n,m} \equiv 2\pi \langle m|\hat{\Pi}(0)|n\rangle, \qquad (6.63)$$

and obtain

$$\text{pr}(\varphi) = \frac{1}{2\pi} \sum_{n,m=0}^{\infty} B_{n,m} \exp[\mathrm{i}(m-n)\varphi]\langle n|\hat{\rho}|m\rangle. \qquad (6.64)$$

Because the operator $\hat{\Pi}(\varphi)$ is Hermitian, the matrix $B_{n,m}$ must be Hermitian as well

$$B_{n,m} = B_{m,n}^{*}. \qquad (6.65)$$

Expressions of the type (6.64) have been known for several phase-dependent distributions for a long time (compare, for instance, Ref. [16]). As we have seen here, the root of these formulas lies in the phase-shifter axiom (A). Now we consider the consequences of axiom (B)

$$\text{pr}(\varphi) = \frac{1}{2\pi} \sum_{n,m=0}^{\infty} B_{n,m} \exp[\mathrm{i}(m-n)\varphi]\langle n|\hat{E}\hat{\rho}\hat{E}^{\dagger}|m\rangle$$

$$= \frac{1}{2\pi} \sum_{n,m=1}^{\infty} B_{n,m} \exp[\mathrm{i}(m-n)\varphi]\langle n-1|\hat{\rho}|m-1\rangle$$

$$= \frac{1}{2\pi} \sum_{n,m=0}^{\infty} B_{n+1,m+1} \exp[\mathrm{i}(m-n)\varphi]\langle n|\hat{\rho}|m\rangle. \qquad (6.66)$$

Consequently, the B coefficients should obey the number-shift invariance as well

$$B_{n+1,m+1} = B_{n,m}. \qquad (6.67)$$

This simple relation will provide us with the key for relating canonical and measured phase distributions.

Because of the invariance relation (6.67) and the Hermitian condition (6.65), the B coefficients depend on a single row of free parameters

$$B_{n,m} = b_{m-n}, \quad b_\nu = \begin{cases} B_{0,\nu} & \text{for } \nu \geq 0 \\ b_{-\nu}^* & \text{for } \nu < 0. \end{cases} \tag{6.68}$$

These parameters characterize all possible noisy phase distributions satisfying both axioms (A) and (B). Using definition (6.68) and the Fock expansion (6.64) for a phase distribution we find

$$\text{pr}(\varphi) = \sum_{\nu=-\infty}^{+\infty} \exp(i\nu\varphi) b_\nu c_\nu \tag{6.69}$$

with

$$c_\nu = \begin{cases} (2\pi)^{-1} \sum_{n=0}^\infty \langle n|\hat{\rho}|n+\nu\rangle & \text{for } \nu \geq 0 \\ c_{-\nu}^* & \text{for } \nu < 0. \end{cases} \tag{6.70}$$

Here the noisy phase distribution $\text{pr}(\varphi)$ is expressed as a Fourier series. According to the convolution theorem we obtain

$$\text{pr}(\varphi) = \frac{1}{2\pi} \int_{-\pi}^{+\pi} g(\phi) \, \text{pr}_p(\varphi - \phi) \, d\phi \tag{6.71}$$

with

$$g(\phi) \equiv \sum_{\nu=-\infty}^{+\infty} \exp(i\nu\phi) b_\nu \tag{6.72}$$

and

$$\text{pr}_p(\varphi) \equiv \sum_{\nu=-\infty}^{+\infty} \exp(i\nu\varphi) c_\nu = \frac{1}{2\pi} \sum_{n,m=0}^\infty \exp[i(m-n)\varphi]\langle n|\hat{\rho}|m\rangle. \tag{6.73}$$

We call $\text{pr}_p(\varphi)$ a *pure canonical phase distribution*. (We remark that $\text{pr}_p(\varphi)$ is the Helstrom–Shapiro–Shepard phase distribution [111], [254] and the Pegg–Barnett phase distribution [19], [22], [213], [214] for a physical state in the infinite-dimensional limit.) The function $g(\phi)$ is real because of the definition (6.68) of the b coefficients. Moreover, it must be nonnegative and normalized to unity because the distributions $\text{pr}(\varphi)$ and $\text{pr}_p(\varphi)$ are nonnegative and normalized for all states. Hence we can interpret $g(\phi)$ as a probability distribution. Our result (6.71) thus means that any noisy phase distribution $\text{pr}(\varphi)$ satisfying both axioms (A) and (B) consists of pure canonical phase distributions $\text{pr}_p(\varphi)$ averaged with respect to a certain probability distribution $g(\phi)$ of reference phases ϕ, which represents the noise.

Finally, it remains to be proved that the pure canonical phase distributions $\text{pr}_p(\varphi)$ are the only ones that deserve the designation "pure," in the sense that they correspond to pure states $\hat{\Pi}(\varphi) = |\varphi\rangle\langle\varphi|$. In fact, they are the only distributions having both properties (A) and (B) and a coefficient matrix $B_{n,m}$ that factorizes according to

$$B_{n,m} = B_n^* B_m. \tag{6.74}$$

The proof is simple. Because of the invariance principle (6.67) we have

$$B_{n+\nu}^* B_{m+\nu} = B_n^* B_m. \tag{6.75}$$

Setting $n = m$ we obtain $|B_n|^2 = |B_0|^2$, and because of the normalization of the phase distribution

$$\int_{-\pi}^{+\pi} \hat{\Pi}(\varphi)\, d\varphi = \sum_{n=0}^{\infty} |B_n|^2 \langle n|\hat{\rho}|n\rangle = |B_0|^2 \sum_{n=0}^{\infty} \langle n|\hat{\rho}|n\rangle = |B_0|^2$$

$$= 1. \tag{6.76}$$

We express B_n as $\exp(-i\beta_n)$ and obtain

$$B_n^* B_1 = \exp[i(\beta_n - \beta_1)] = B_{n-1}^* B_0 = \exp[i(\beta_{n-1} - \beta_0)] \tag{6.77}$$

and, consequently,

$$\exp(i\beta_n) = \exp(i\beta_{n-1}) \exp(i\phi), \qquad \phi = \beta_1 - \beta_0. \tag{6.78}$$

Applying this relation n times we get finally

$$B_n = \exp(-i\beta_n) = \exp(-i\beta_0) \exp(-in\phi). \tag{6.79}$$

Hence the phase distribution $\mathrm{pr}(\varphi)$ is

$$\mathrm{pr}(\varphi) = \frac{1}{2\pi} \sum_{n,m=0}^{\infty} \exp[i(m-n)(\varphi - \phi)]\langle n|\hat{\rho}|m\rangle$$

$$= \mathrm{pr}_p(\varphi - \phi). \tag{6.80}$$

Note that this distribution was first introduced by London [177] prior to Dirac's paper on the field quantization via number and phase. We see that up to a reference phase our basic axioms (A) and (B) determine a canonical phase distribution uniquely when we consider a pure distribution, and, in general, any noisy phase distribution satisfying the complementary axioms can be regarded as a statistical mixture of pure canonical phase distributions.

6.3.2 Measured phase distribution

We have seen that the phase distribution measured in the Noh–Fougères–Mandel experiment is the radius-integrated Q function (Q phase), provided a strong local oscillator and perfect detectors have been employed. Also, in other operational approaches to quantum phase [16], [253], the Q-phase distribution is measured. In this section we show that we can interpret this operational phase distribution as a smoothed canonical phase distribution. The Q-phase distribution reads in terms of the POM formalism

$$\mathrm{pr}_Q(\varphi) = \mathrm{tr}\{\hat{\rho}\hat{\Pi}_Q(\varphi)\} \tag{6.81}$$

with

$$\hat{\Pi}_Q(\varphi) \equiv \frac{1}{\pi} \int_0^\infty r|r\exp(\mathrm{i}\varphi)\rangle\langle r\exp(\mathrm{i}\varphi)|\,dr. \tag{6.82}$$

Here $|r\exp(\mathrm{i}\varphi)\rangle$ denotes a coherent state of amplitude r and phase φ. (This type of phase state for the Q phase has been studied in detail by Paul [210].) Does the Q phase obey the requirements for a canonical phase? To decide, we calculate the Fock representation of the $\hat{\Pi}_Q(\varphi)$ operator

$$\langle m|\hat{\Pi}_Q(\varphi)|n\rangle = \frac{1}{\pi} \int_0^\infty \langle m|r\exp(\mathrm{i}\varphi)\rangle\langle r\exp(\mathrm{i}\varphi)|n\rangle r\,dr$$

$$= \frac{1}{\pi} \exp[\mathrm{i}(m-n)\varphi]\frac{1}{\sqrt{m!n!}} \int_0^\infty \exp(-r^2)r^{m+n+1}\,dr \tag{6.83}$$

using the Fock expansion (2.63) of coherent states. The radius integral gives $1/2[(n+m)/2]!$ (see, for instance, Ref. [225], Vol. I, Eq. 2.3.3.1), where we denote the Gamma function $\Gamma(x+1)$ by the generalized factorial $x!$. Consequently, we obtain in the Fock representation

$$\mathrm{pr}_Q(\varphi) = \frac{1}{2\pi} \sum_{n,m=0}^\infty B_{n,m} \exp[\mathrm{i}(m-n)\varphi]\langle n|\hat{\rho}|m\rangle \tag{6.84}$$

with

$$B_{n,m} = \frac{[(n+m)/2]!}{(n!m!)^{1/2}}. \tag{6.85}$$

Obviously, $\mathrm{pr}_Q(\varphi)$ satisfies axiom (A). Moreover, it shows the number-shift invariance (B) to a good approximation

$$B_{n+1,m+1} = \frac{\frac{1}{2}(n+1+m+1)}{(n+1)^{1/2}(m+1)^{1/2}} B_{n,m} \approx B_{n,m}. \tag{6.86}$$

The B coefficients differ by the ratio of the arithmetic and geometric mean of the photon numbers $n+1$ and $m+1$. When n and m are large, the invariance principle (6.67) is fulfilled quite well. This means that the Q-phase distribution is approximately a canonical phase distribution. On the other hand, we note that the B coefficients tend to unity in the limit $n,m \to \infty$. Thus, the Q phase coincides with the pure canonical phase for very large photon numbers. Hence, the number-shift invariance cannot be exact. Here we are interested in an intermediate regime where n and m are relatively large.

What is the asymptotics of the Q-phase distribution? For relatively narrow photon-number distributions compared to the mean, we can approximate

$$B_{n,n+|\nu|} \approx B_{N,N+|\nu|} \tag{6.87}$$

with N being the mean photon number

$$N = \text{tr}\{\hat{\rho}\hat{a}^\dagger\hat{a}\}. \tag{6.88}$$

Consequently, the Q-phase distribution reads

$$\text{pr}_Q(\varphi) = \sum_{\nu=-\infty}^{+\infty} \exp(i\nu\varphi)B_{N,N+|\nu|}c_\nu \tag{6.89}$$

where the c_ν coefficients are defined in Eq. (6.70). Similarly to noisy canonical phase distributions treated in the previous section, we obtain according to the convolution theorem

$$\text{pr}_Q(\varphi) = \frac{1}{2\pi} \int_{-\pi}^{+\pi} g(\phi; N)\,\text{pr}_p(\varphi - \phi)\,d\phi \tag{6.90}$$

with

$$g(\phi; N) \equiv \sum_{\nu=-\infty}^{+\infty} \exp(i\nu\phi)B_{N,N+|\nu|}. \tag{6.91}$$

Note that the mean photon number N enters the expression (6.91) for the reference-phase distribution $g(\phi; N)$. We wish to derive an asymptotic expression for $g(\phi; N)$ in the case of a large N. We replace the Fourier series in Eq. (6.91) by an integral and use the saddle-point method to evaluate it. Applying the improved Stirling formula [173]

$$x! \sim (2\pi)^{1/2}\left(x + \frac{1}{2}\right)^{x+1/2} \exp\left(-x - \frac{1}{2}\right) \tag{6.92}$$

we obtain from Eq. (6.85)

$$\begin{aligned}
\ln B_{N,N+|\nu|} &\sim \left(N + \frac{|\nu|}{2} + \frac{1}{2}\right) \ln\left(N + \frac{|\nu|}{2} + \frac{1}{2}\right) \\
&\quad - \frac{1}{2}\left(N + \frac{1}{2}\right) \ln\left(N + \frac{1}{2}\right) \\
&\quad - \frac{1}{2}\left(N + |\nu| + \frac{1}{2}\right) \ln\left(N + |\nu| + \frac{1}{2}\right) \\
&= -\frac{\nu^2}{8(N + 1/2)} + \mathcal{O}(|\nu|^3). \tag{6.93}
\end{aligned}$$

Hence, we get for the distribution function

$$g(\phi; N) = \int_{-\infty}^{+\infty} \exp\left[i\nu\phi - \frac{\nu^2}{8(N + 1/2)}\right] d\nu \tag{6.94}$$

and, finally,

$$g(\phi; N) = 2\pi \left[\frac{2(N + 1/2)}{\pi} \right]^{1/2} \exp\left[-2\left(N + \frac{1}{2} \right)\phi^2 \right]. \tag{6.95}$$

This simple formula describes the asymptotics of the reference-phase distribution $g(\phi; N)$. It is a Gaussian distribution with a width depending inversely on the mean photon number N (plus the vacuum term $1/2$), which reflects the extra noise involved in realistic phase measurements. The distribution $g(\phi; N)$ gets narrower with increasing N and finally tends to a δ-function for very large N, because with increasing intensity the influence of the noise decreases. In the macroscopic regime the measurement-induced noise is negligible. This is readily understood from the particular source of extra noise being present in an experimental setup that allows measurement of the Q phase. It is the vacuum noise intruding the apparatus via the unused port of a beam splitter (or the amplification noise in the Bandilla–Paul scheme [16]). In any case, the noise becomes negligible when the intensity of the initial field is high.

As the final result of our analysis, we have bridged two different concepts of the quantum-optical phase – the canonical phase and the operational approach. We have seen that the operationally defined (the measured) phase is a noisy canonical phase. The reason is that the Noh–Fougères–Mandel experiment is essentially a simultaneous measurement of number and phase. Moreover, we have quantified this extra noise in the semiclassical regime. The noise depends inversely on the intensity and vanishes when we leave the "quantum sector" and reach the domain of classical optics.

6.4 Further reading

The original Gedanken experiment of Arthurs and Kelly is reviewed in a beautiful paper [259] by S. Stenholm. Note that Arthurs and Kelly studied a case in which three quantum objects were involved: the system and two meters. In particular, the system is not lost after the detection process but prepared in a state governed by the measurement results. So strictly speaking, the eight-port homodyne detector is not a literal realization of the original idea in quantum optics because here the signal is split, guided to the two homodyne detectors, and then, of course, destroyed. A quantum-optical scheme that comes closer to the Gedanken experiment of Arthurs and Kelly was proposed by P. Törmä, S. Stenholm, and I. Jex [267]. A physically simple example of a joint measurement of position and momentum was considered by M.G. Raymer [228].

S.L. Braunstein, C.M. Caves, and G.J. Milburn [40] and G.S. Agarwal and S. Chaturvedi [4] studied generalizations of the Arthurs and Kelly scheme or of the eight-port homodyne detector, respectively, to measure the positive P representation of P.D. Drummond and C.W. Gardiner [81]. (The positive P representation is another quasiprobability distribution and has the merit of being always nonnegative but involves, however, a doubling of the phase space.)

For the theory of multiport homodyning, see the early yet excellent paper by N.G. Walker [283]. Recent work was done by H. Kühn, D.-G. Welsch, and W. Vogel [141] and A. Zucchetti, W. Vogel, M. Tasche, and D.-G. Welsch [307]. For the violation of Bell's inequality in eight-port homodyne detection see Ref. [164].

Propensities, that is, phase-space distributions that are convolutions of the Wigner function with a given ruler Wigner function, are studied in detail in a paper [144] by D. Lalović, D.M. Davidović, and N. Bijedić. V. Bužek, C.H. Keitel, and P.L. Knight [48], [49] introduced and examined sampling entropies associated with operational phase-space measurements. B.-G. Englert and K. Wódkiewicz [86] propagated the notion of operational and intrinsic observables in quantum mechanics. Many aspects of joint measurements of conjugate variables are also considered in the book [46] by P. Busch, M. Grabowski, and P.J. Lahti.

The operational approach to the quantum-optical phase by J.W. Noh, A. Fougères, and L. Mandel [197]–[201] initiated a series of papers. In particular, the relation between phase measurements and the Q function for strong local oscillators was pointed out by M. Freyberger, K. Vogel, and W.P. Schleich [96], [97] and in Ref. [152]. See also the articles by G.M. D'Ariano and M.G.A. Paris [66], [67]; A. Lukš, V. Peřinová, and J. Křepelka [184]; P. Busch, M. Grabowski, and P.J. Lahti [45]; and B.-G. Englert, K. Wódkiewicz, and P. Riegler [87], [237]. H.M. Wiseman [292] showed how to go beyond the Q phase in adaptive phase measurements. The concept of the quantum-optical phase is still a subject of debate. See for instance the paper [266] by J.R. Torgerson and L. Mandel.

Although the canonical phase cannot be easily measured, it can be reconstructed from the quadrature distributions measured in homodyne detection; see the experiments [24], [25], [256], [257] by M. Beck, D.T. Smithey, J. Cooper, and M.G. Raymer.

7

Summary

Measuring the quantum state of light – this was the theme that guided us on a journey through a part of modern quantum optics. We started with some considerations and speculations about the nature of states in physics and in particular in quantum physics. In a brief chapter about the *quantum theory of light* we introduced the concept of the electromagnetic oscillator (the single-mode description) and studied a number of typical single-mode states. (In a later section on *spatial–temporal modes* we returned to the electromagnetic oscillator and showed how these modes are selected from the light field.) In the chapter about *quasiprobability distributions* we motivated the idea of the phase space in quantum mechanics and in particular the Wigner representation, following an axiomatic approach. We derived the basic properties of the Wigner function and illustrated this quantum phase-space description by examples. Additionally, we introduced the Q and the P functions and, finally, the class of s-parameterized quasiprobability distributions. Another chapter was devoted to the quantum theory of *simple optical instruments*. We have seen that such an innocent device like a beam splitter may show some intriguing quantum effects when confronted with the quantum nature of light. After examining various aspects of beam splitters, we turned to the theory of balanced homodyne detection. As we have seen, this detection scheme is a highly efficient experimental technique to measure the quantum properties of light.

These preparatory chapters equipped us for approaching the core of this book – *quantum tomography* (the only chapter where the number of equations exceeds the magic value of 99). First, we studied *phase-space tomography*. In particular, we introduced the basics of tomography (the inverse Radon transformation), sketched the idea of the numerical implementation (filtered backprojection algorithm), and, finally, discussed how quantum mechanics can be formulated without probability amplitudes. We studied *quantum-state sampling* as an alternative way of doing or understanding phase-space tomography.

As the central theoretical tool we introduced the pattern functions and examined their properties with the help of a theorem on the Schrödinger equation. We provided ready-to-use instructions for the required data processing (fast and stable algorithms to compute the pattern functions and to calculate the Wigner function from the sampled density matrix). *How precisely can we measure quantum states?* We estimated the effects of a finite statistical ensemble, of a limited number of reference phases, of a finite bin width, and, of course, of detection losses.

In view of Heisenberg's uncertainty principle the *simultaneous measurement of position and momentum* sounds paradoxical. However, we have seen that we can "circumvent" the uncertainty principle by taking it literally, that is, by measuring the position and the momentum jointly yet not precisely. One effect of this approach was a doubling of the uncertainty product, for instance. We studied an interesting quantum-optical device, the eight-port homodyne detector, where this idea became experimental reality. In particular, we calculated the measured phase-space density and related it to the quantum-mechanical quasiprobability distributions. Finally, we studied how this device measures the quantum-optical phase. We defined axiomatically the canonical phase distribution and related it to the measured phase in the semiclassical regime.

Measuring the quantum state of light and, more generally, state reconstruction in quantum mechanics is a field still growing and diversifying. New ideas are appearing, and only time (not this book) can select the most successful concepts. Nobody is wise enough to predict the future of a "newborn baby." To avoid misjudgments about present and future work, the very latest results "at the forefront of this rapidly changing field" are not mentioned. With my apologies I leave the reader to the current literature. Hopefully, after having read this book, the reader will be armed with the theoretical tools and the right questions to join this enterprise. Especially experimentalists are welcome to apply the sketched ideas in new and exciting investigations into the affairs of the quanta.

7.1 Acknowledgments

Most of the text was written during a wonderful year at the University of Oregon, where I enjoyed collaborating with Michael G. Raymer. I am grateful for our countless conversations and for his careful reading of the manuscript. From him I gained a deep respect for the art of experimentation (the patient "digging for gold").

My personal contribution to the story of quantum-state measurements began when I joined the *Arbeitsgruppe Nichtklassische Strahlung der Max-Planck-Gesellschaft an der Humboldt-Universität zu Berlin* to work with Harry Paul.

I have been very fortunate to learn from him the art of theoretical quantum optics. Moreover, I enjoyed his noble, generous support.

This book would not exist without Irina. I thank her for her constant encouragement and help.

I am grateful to Gerd Breitenbach and Michael Munroe for our collaboration and especially for providing me with their impressive figures.

Many people contributed to the book by giving valuable comments on the text or, indirectly, by discussions and support. I thank in particular D. Adkison, H.A. Bachor, U. Bandelow, A. Bandilla, S.M. Barnett, B. Böhmer, S.L. Braunstein, V. Bužek, H.J. Carmichael, C.M. Caves, G.M. D'Ariano, A. Faridani, M. Freyberger, U. Herzog, Z. Hradil, U. Janicke, I. Jex, T. Kiss, P.L. Knight, L. Knöll, D.S. Krähmer, J.-P. Kuska, J. Lehner, R. Loudon, D.F. McAlister, B. Mecking, M. Mlejnek, J. Mlynek, T.W. Mossberg, C. Müller, T. Opatrný, A. Orlowski, S.F. Pereira, Th. Richter, V.I. Savichev, S. Schiller, W.P. Schleich, S. Stenholm, P. Törmä, J.A. Vaccaro, K. Vogel, W. Vogel, I.A. Walmsley, H. Walther, D.-G. Welsch, M. Wilkens, A. Wünsche, and P. Zoller.

I thank the Gilkey and Hendrickson families for making our stay in Eugene so enjoyable.

The work was financially supported by an Otto Hahn Award of the Max Planck Society and by a Habilitation Fellowship of the Deutsche Forschungsgemeinschaft.

Appendix 1

Semiclassical approximation

In this appendix we sketch the elements of a semiclassical theory for the regular and the irregular wave functions. The technique we use is known as the Wentzel–Kramers–Brillouin (WKB) method. Originally [43], [139], [288], it was developed to understand how the "old quantum theory" of Bohr and Sommerfeld is related to the "new theory" of Schrödinger. The WKB method is a powerful tool for finding accurate analytical approximations for complicated problems in quantum mechanics. Because the method is semiclassical, it is also helpful for understanding the underlying physics. However, this appendix is not the place to develop explicit proofs of the WKB technique. (The reader is referred to two textbooks [34], [145], for instance.) We describe the main results only briefly.

The semiclassical solution of the Schrödinger equation, with the potential $U(x)$,

$$\left[-\frac{1}{2}\frac{\partial^2}{\partial x^2} + U(x)\right]\phi_n = \omega_n\phi_n \tag{A1.1}$$

is given by the WKB formula [145]

$$\phi_n = c_n^{(1)}p_n^{-1/2}\exp(iS_n) + c_n^{(2)}p_n^{-1/2}\exp(-iS_n). \tag{A1.2}$$

Here

$$p_n(x) = \sqrt{2\omega_n - 2U(x)} \tag{A1.3}$$

denotes the classical momentum associated with the total energy ω_n. It vanishes at the left and the right turning points l_n and r_n of the classical motion

$$p_n(l_n) = p_n(r_n) = 0. \tag{A1.4}$$

The region outside (l_n, r_n) is classically forbidden because there the momentum

$p_n(x)$ becomes purely imaginary. The quantity

$$S_n(x) = \int_{r_n}^{x} p_n(x')\,dx' \tag{A1.5}$$

is often called the time-independent part of the classical action [145]. In the classically forbidden zone $S_n(x)$ is purely imaginary as well, implying that in this region one part of the semiclassical solution (A1.2) becomes exponentially small, whereas the other grows exponentially large. We will associate the decreasing part with a regular wave function ψ_n and the increasing part with an irregular solution φ_n. The energy ω_n is given by the famous Bohr–Sommerfeld quantization rule

$$\oint p_n\,dx = 2\pi\left(n + \frac{1}{2}\right). \tag{A1.6}$$

The semiclassical solution (A1.2) is valid for large energies ω_n compared with unity (since $\hbar = 1$), that is, for high quantum numbers. We note that the approximation (A1.2) is no longer justified near the classical turning points r_n (in the Bohr–Sommerfeld band [79]) because there $p_n^{-1/2}$ diverges. The Bohr–Sommerfeld band [79] separates the regions where the quasiclassical theory is valid. However, by going to the complex plane we could link the semiclassical solution (A1.2) in the classically allowed region with the solution (A1.2) in the forbidden zone [145]. Alternatively, we may linearize the potential $U(x)$ in the Bohr–Sommerfeld band near the classical turning point and match the WKB wave functions outside. We approximate Eq. (A1.1) by the stationary Schrödinger equation

$$\left[-\frac{1}{2}\frac{\partial^2}{\partial x^2} - F_n(x - r_n)\right]\phi_n = 0 \tag{A1.7}$$

for a one-dimensional wave packet driven by the constant force

$$F_n = -\left.\frac{dU}{dx}\right|_{r_n}. \tag{A1.8}$$

We can solve the approximate Schrödinger equation (A1.7) exactly and obtain two fundamental solutions:

$$\psi_n = c_n\sqrt{\pi}\,|2F_n|^{-1/6}\mathrm{Ai}[|2F_n|^{1/3}(x - r_n)], \tag{A1.9}$$

$$\varphi_n = 2c_n^{-1}\sqrt{\pi}\,|2F_n|^{-1/6}\mathrm{Bi}[|2F_n|^{1/3}(x - r_n)]. \tag{A1.10}$$

Here Ai and Bi denote the Airy functions as defined in Ref. [2]. We note that the prefactors $\sqrt{\pi}\,|2F_n|^{-1/6}$ were chosen to obtain convenient expressions for the matched WKB wave functions. Additionally, the solutions (A1.9) and (A1.10) are designed such that the Wronskian $\psi_n\varphi_n' - \psi_n'\varphi_n$ equals 2. [To verify this property see Ref. [2], Eq. 10.4.10.] Let us match the wave functions ψ_n and φ_n

with the WKB expressions (A1.2) in the classically allowed and in the forbidden zone near the Bohr–Sommerfeld band. Here the momentum (A1.3) becomes

$$p_n(x) = \sqrt{|F_n|(r_n - x)},\tag{A1.11}$$

whereas the action (A1.5) tends to

$$S_n(x) = \frac{2}{3}|F_n|^{1/2}(r_n - x)^{3/2}.\tag{A1.12}$$

Inserted in Eq. (A1.2) these expressions describe how the WKB wave functions behave near the Bohr–Sommerfeld band. To find the match with the solutions (A1.9) and (A1.10), we use the known asymptotic properties of the Airy functions. See Ref. [2], Eqs. 10.4.59, 10.4.60, 10.4.63, and 10.4.64. We find that an evanescent wave function in the classically forbidden zone $x > r_n$

$$\psi_n = \frac{c_n}{2}|p_n|^{-1/2}\exp(-|S_n|)\tag{A1.13}$$

approaches the first fundamental solution (A1.9) in the Bohr–Sommerfeld band. Then this wave function matches a standing wave in the classically allowed region

$$\psi_n = c_n p_n^{-1/2}\cos\left(S_n + \frac{\pi}{4}\right).\tag{A1.14}$$

Using the same arguments for the second solution, we find that an exponentially growing wave function in the forbidden zone

$$\varphi_n = 2c_n^{-1}|p_n|^{-1/2}\exp(+|S_n|)\tag{A1.15}$$

is similarly linked with (A1.10) and a standing wave inside the classically allowed region

$$\varphi_n = 2c_n^{-1}p_n^{-1/2}\sin\left(S_n + \frac{\pi}{4}\right).\tag{A1.16}$$

This wave is, however, out of phase compared with the regular wave function ψ_n. Finally, we note that for all given semiclassical approximations the Wronskian $\psi_n\varphi_n' - \psi_n'\varphi_n$ equals exactly 2. In view of this fact we can rightfully regard the expressions (A1.10), (A1.15), and (A1.16) as semiclassical formulas for the irregular wave functions.

So far we have sketched the general WKB technique and extended it to irregular wave functions for arbitrary potentials. In this book we are mainly interested in light modes, that is, in harmonic oscillators with the potential

$$U(x) = \frac{x^2}{2}\tag{A1.17}$$

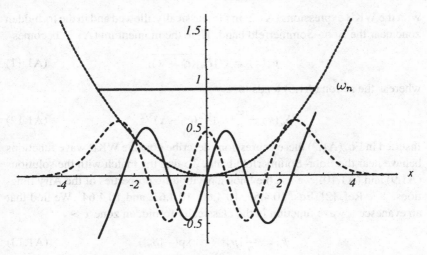

Fig. A1.1. Semiclassical wave functions for harmonic potential (dotted line). Classically, a particle with given energy ω_n would be confined in the region $l_n \leq x \leq r_n$ [with $l_n = -r_n$ and r_n given by Eq. (A1.19) for the harmonic oscillator]. Quantum-mechanically, the particle is represented by a standing wave (dashed line) in the classically allowed region. This wave decays exponentially in the classically forbidden zone. An unnormalizable solution of the Schrödinger equation, that is, an irregular wave function (solid line), grows exponentially in the forbidden zone. In the classically allowed region it behaves like a standing wave as well, which is, however, out of phase compared to the regular wave function.

in our units. For harmonic oscillators the semiclassical energy ω_n agrees with the exact result

$$\omega_n = n + \frac{1}{2}. \tag{A1.18}$$

The classical turning point r_n (the Bohr–Sommerfeld radius) is given by

$$r_n = \sqrt{2n + 1}. \tag{A1.19}$$

In the classically allowed region $|x| < r_n$ we may parameterize x and p_n by

$$x = r_n \cos t_n \tag{A1.20}$$

and

$$p_n = r_n \sin t_n \tag{A1.21}$$

to obtain the action S_n according to Eq. (A1.5)

$$S_n = \frac{r_n^2}{4}[\sin(2t_n) - 2t_n]. \tag{A1.22}$$

In the forbidden zone $x > r_n$ we parameterize x and p_n by

$$x = r_n \cosh t_n \qquad (A1.23)$$

and

$$p_n = i r_n \sinh t_n \qquad (A1.24)$$

(with a different t_n) and get the purely imaginary action

$$S_n = i \frac{r_n^2}{4} [\sinh(2t_n) - 2t_n]. \qquad (A1.25)$$

The force F_n in the Bohr–Sommerfeld band $x \approx r_n$ is given by

$$F_n = -r_n. \qquad (A1.26)$$

Finally, we must work out the prefactors in our expressions (A1.9–A1.10) and (A1.13–A1.16). For this step we simply match the asymptotic expression (5.85) of the exact regular wave function with the semiclassical solution (A1.13) in the forbidden zone. Here the wave function (A1.13) is approximately

$$\psi_n(x) \sim \frac{c_n}{2\sqrt{x}} \exp\left[-\frac{x^2}{2} + \frac{r_n^2}{4} + \frac{r_n^2}{2} \ln\left(\frac{2x}{r_n}\right)\right] \qquad (A1.27)$$

or, expressed differently,

$$\psi_n(x) \sim c_n 2^{n/2-3/4} \left(n + \frac{1}{2}\right)^{-(n+1/2)/2} x^n \exp\left[-\frac{x^2}{2} + \frac{1}{2}\left(n + \frac{1}{2}\right)\right]. \qquad (A1.28)$$

We use the improved Stirling formula (6.92) and obtain

$$\psi_n(x) \sim \frac{c_n}{\sqrt{2}} \pi^{1/4} \left(\frac{2^n}{n!}\right)^{1/2} x^n \exp\left(-\frac{x^2}{2}\right), \qquad (A1.29)$$

which implies

$$c_n = \sqrt{\frac{2}{\pi}} \qquad (A1.30)$$

when compared with Eq. (5.85).

We note that the regular WKB approximations are known as the Plancherel–Rotach formulas for Hermite polynomials [262].

Appendix 2

A theorem on the Schrödinger equation

In this appendix we prove the following theorem [168, 235] on the Schrödinger equation (A1.1): Suppose that $\psi_n(x)$ and $\varphi_n(x)$ are the regular and the irregular solutions of the equation (A1.1) for a given value of ω_n. Then the integral

$$G^{mn}_{\mu\nu} \equiv \int_{-\infty}^{+\infty} \psi_\mu \psi_\nu \partial(\psi_m \varphi_n) \, dx \qquad (A2.1)$$

is equal to the product of the Kronecker symbols $\delta_{\mu m}$ and $\delta_{\nu n}$ if the frequencies fulfill the condition

$$\omega_\mu - \omega_\nu = \omega_m - \omega_n \qquad (A2.2)$$

and the Wronskian of $\psi_n(x)$ and $\varphi_n(x)$ equals two, that is, if

$$\psi_n \varphi'_n - \psi'_n \varphi_n = 2. \qquad (A2.3)$$

Here we abbreviate the partial derivative with respect to x by ∂ or using a prime.

The proof of this theorem is inspired by the same principal idea as in the classic orthogonality proof for eigenfunctions of Hermitian operators [145, Section 3]. As an additional ingredient, we will persistently use partial integration and take advantage of the Schrödinger equation (A1.1) and of the frequency constraint (A2.2). First, we derive from the stationary Schrödinger equation (A1.1) two Schrödinger-like equations, one for the product of the wave functions

$$\partial^2(\psi_\mu \psi_\nu) = 2U_{\mu\nu} \psi_\mu \psi_\nu + 2\psi'_\mu \psi'_\nu \qquad (A2.4)$$

and another for $\partial(\psi_m \varphi_n)$

$$\partial^3(\psi_m \varphi_n) = 4U_{mn} \partial(\psi_m \varphi_n) + 4U' \psi_m \varphi_n - 2(\omega_m - \omega_n)W_{mn} \qquad (A2.5)$$

with the modified potential

$$U_{mn} \equiv 2U - \omega_m - \omega_n. \qquad (A2.6)$$

180

We introduce the generalized Wronskians

$$W_{mn} \equiv \psi_m \varphi'_n - \psi'_m \varphi_n \qquad (A2.7)$$

and

$$V_{\mu\nu} \equiv \psi_\mu \psi'_\nu - \psi'_\mu \psi_\nu. \qquad (A2.8)$$

It is easy to see from the Schrödinger equation (A1.1) that the derivatives of the Wronskians W_{mn} and $V_{\mu\nu}$ obey the relations

$$W'_{mn} = 2(\omega_m - \omega_n)\psi_m \varphi_n \qquad (A2.9)$$

and

$$V'_{\mu\nu} = 2(\omega_\mu - \omega_\nu)\psi_\mu \psi_\nu. \qquad (A2.10)$$

To prove the orthogonality of $\psi_\mu \psi_\nu$ and $\partial(\psi_m \varphi_n)$ we replace in the scalar product (A2.1) the product of the wave functions $\psi_\mu \psi_\nu$ using the first Schrödinger-like equation (A2.4)

$$2(\omega_\mu + \omega_\nu)G_{\mu\nu}^{mn} = \int_{-\infty}^{+\infty} [4U\psi_\mu \psi_\nu - \partial^2(\psi_\mu \psi_\nu) + 2\psi'_\mu \psi'_\nu]\partial(\psi_m \varphi_n)\,dx. \qquad (A2.11)$$

Then we integrate by parts to move the differential operator ∂^2 to the sampling function $\partial(\psi_m \varphi_n)$ and apply the second Schrödinger-like equation (A2.5). In this way we obtain

$$2(\omega_\mu + \omega_\nu)G_{\mu\nu}^{mn} = \int_{-\infty}^{+\infty} \psi_\mu \psi_\nu(4U\partial - 4U_{mn}\partial - 4U')\psi_m \varphi_n\,dx + H_{\mu\nu}^{mn} \qquad (A2.12)$$

with the additional term

$$H_{\mu\nu}^{mn} = 2\int_{-\infty}^{+\infty} [\psi'_\mu \psi'_\nu \partial(\psi_m \varphi_n) + (\omega_m - \omega_n)\psi_\mu \psi_\nu W_{mn}]\,dx. \qquad (A2.13)$$

To evaluate $H_{\mu\nu}^{mn}$ we integrate $\psi'_\mu \psi'_\nu \partial(\psi_m \varphi_n)$ by parts and we use the frequency constraint (A2.2) and the relation (A2.10) of the generalized Wronskian $V_{\mu\nu}$ to arrive at the intermediate result

$$H_{\mu\nu}^{mn} = \int_{-\infty}^{+\infty} [-2\psi_m \varphi_n \partial(\psi'_\mu \psi'_\nu) + V'_{\mu\nu} W_{mn}]\,dx. \qquad (A2.14)$$

We integrate $V'_{\mu\nu} W_{mn}$ by parts and use

$$-\partial(\psi'_\mu \psi'_\nu) = -U_{\mu\nu}\partial(\psi_\mu \psi_\nu) + (\omega_\mu - \omega_\nu)V_{\mu\nu} \qquad (A2.15)$$

and again the frequency constraint (A2.2) together with relation (A2.9) of the

Wronskian W_{mn} to obtain the final result

$$H_{\mu\nu}^{mn} = -2 \int_{-\infty}^{+\infty} \psi_m \varphi_n U_{\mu\nu} \partial(\psi_\mu \psi_\nu) \, dx$$

$$= +2 \int_{-\infty}^{+\infty} \psi_\mu \psi_\nu (U_{\mu\nu} \partial + 2U') \psi_m \varphi_n \, dx. \qquad (A2.16)$$

We insert this expression for $H_{\mu\nu}^{mn}$ in Eq. (A2.12) and see immediately from definition (A2.6) that

$$2(\omega_m + \omega_n - \omega_\mu - \omega_\nu) G_{\mu\nu}^{mn} = 0. \qquad (A2.17)$$

This means that if the sums $\omega_m + \omega_n$ and $\omega_\mu + \omega_\nu$ are not equal, $G_{\mu\nu}^{mn}$ must vanish. Because the differences $\omega_m - \omega_n$ and $\omega_\mu - \omega_\nu$ coincide according to the frequency constraint (A2.2), $G_{\mu\nu}^{mn}$ must be proportional to $\delta_{\mu m}$ and $\delta_{\nu n}$.

What happens when $m = \mu$ and $n = \nu$, that is, when $\omega_m + \omega_n - \omega_\mu - \omega_\nu = 0$? First, we note that according to Eq. (A2.9) the Wronskian W_{nn} is a constant (as it is well known for solutions of the Schrödinger equation.) We differentiate $\psi_m \varphi_n$ in G_{mn}^{mn} and use the Wronskian and the normalization of the regular eigenfunction ψ_m to obtain

$$G_{mn}^{mn} = \int_{-\infty}^{+\infty} \psi_m \varphi_n \partial(\psi_m \psi_n) \, dx + W_{nn}. \qquad (A2.18)$$

On the other hand, partial integration in the definition (A2.1) of G_{mn}^{mn} gives

$$G_{mn}^{mn} = - \int_{-\infty}^{+\infty} \psi_m \varphi_n \partial(\psi_m \psi_n) \, dx. \qquad (A2.19)$$

Adding Eqs. (A2.18) and (A2.19) implies that $2G_{mn}^{mn}$ must equal the Wronskian W_{nn}. So if we require the Wronskian condition (A2.3) the theorem is proven.

Because we have used partial integration in the proof, the theorem is valid if quantities such as $\psi_m \varphi_n U_{\mu\nu} \psi_\mu \psi_\nu$ decay fast enough at the boundaries $-\infty$ and $+\infty$. That this is correct can be seen from the known asymptotic behavior of one-dimensional wave packets. In the classically forbidden zone the semi-classical approximation becomes perfect when x tends to infinity. The product of the regular and the irregular wave functions approaches

$$\psi_m \varphi_n \sim c_m c_n^{-1} |p_n p_m|^{-1/2} \exp(|S_n| - |S_m|) \qquad (A2.20)$$

with

$$|S_n| - |S_m| = \int_{r_n}^{x} [|p_n(\xi)| - |p_m(\xi)|] \, d\xi + c$$

$$= \int_{r_n}^{x} \frac{2(\omega_m - \omega_n)}{|p_n(\xi)| + |p_m(\xi)|} \, d\xi + c. \qquad (A2.21)$$

[Because the energies ω_n and ω_m are different in general the classical turning point r_n differs from r_m as well. Therefore we obtain a constant c in Eq. (A2.21).] This expression tends to

$$|S_n| - |S_m| \sim (\omega_m - \omega_n) \int_{r_n}^{x} [2U(\xi)]^{-1/2}\, d\xi + c' \tag{A2.22}$$

for large x [with another (unimportant) constant c'.] We see that for $\omega_n \geq \omega_n$ the product of ψ_m and φ_n decays nicely. On the other hand, if $\omega_n < \omega_n$ the growth of $\partial(\psi_m \varphi_n)$ is compensated for by the exponential decay of the regular wave functions ψ_μ and ψ_ν.

Bibliography

[1]   Abbas, G.L., Chan, V.W.S., and Yee, T.K. (1983), *Opt. Lett.* **8**, 419.
[2]   Abramowitz, M., and Stegun, I.A. (1970), *Handbook of Mathematical Functions* (Dover, New York).
[3]   Agarwal, G.S. (1981), *Phys. Rev. A* **24**, 2889.
[4]   Agarwal, G.S., and Chaturvedi, S. (1994), *Phys. Rev. A* **49**, R665.
[5]   Aharonov, Y., Falkoff, D., Lerner, E., and Pendleton, H. (1966), *Ann. Phys.* (New York) **39**, 498.
[6]   Aharonov, Y., Albert, D.Z., and Au, C.K. (1981), *Phys. Rev. Lett.* **47**, 1029.
[7]   Allen, L., and Stenholm, S. (1992), *Opt. Commun.* **93**, 253.
[8]   Arthurs, E., and Kelly, J.L., Jr. (1965), *Bell Syst. Tech. J.* **44**, 725.
[9]   Bacry, H., Grossmann, A., and Zak, J. (1975), *Phys. Rev. B* **12**, 1118.
[10]   Balazs, N.L., and Jennings, B.K. (1984), *Phys. Rep.* **104**, 347.
[11]   Ballentine, L.E. (1990), *Quantum Mechanics* (Prentice Hall, Englewood Cliffs, N.J.).
[12]   Band, W., and Park, J.L. (1970), *Found. Phys.* **1**, 133.
[13]   Band, W., and Park, J.L. (1971), *Found. Phys.* **1**, 211.
[14]   Band, W., and Park, J.L. (1971), *Found. Phys.* **1**, 339.
[15]   Band, W., and Park, J.L. (1979), *Am. J. Phys.* **47**, 188.
[16]   Bandilla, A., and Paul, H. (1969), *Ann. Phys.* (Leipzig) **23**, 323.
[17]   Bandilla, A., and Paul, H. (1970), *Ann. Phys.* (Leipzig) **24**, 119.
[18]   Barnett, S.M., and Pegg, D.T. (1986), *J. Phys. A* **19**, 3849.
[19]   Barnett, S.M., and Pegg, D.T. (1989), *J. Mod. Opt.* **36**, 7.
[20]   Barnett, S.M., and Phoenix, S.J.D. (1989), *Phys. Rev. A* **40**, 2404.
[21]   Barnett, S.M., and Phoenix, S.J.D. (1991), *Phys. Rev. A* **44**, 535.
[22]   Barnett, S.M., and Pegg, D.T. (1992), *J. Mod. Opt.* **39**, 2121.
[23]   Barnum, H., Caves, C.M., Fuchs, C.A., Josza, R., and Schumacher, B. (1996), *Phys. Rev. Lett.* **76**, 2818.
[24]   Beck, M., Smithey, D.T., Cooper, J., and Raymer, M.G. (1993), *Opt. Lett.* **18**, 1259.
[25]   Beck, M., Smithey, D.T., and Raymer, M.G. (1993), *Phys. Rev. A* **48**, R890.
[26]   Bell, J.S. (1964), *Physics* **1**, 195.
[27]   Bell, J.S. (1987), *Speakable and Unspeakable in Quantum Mechanics* (Cambridge University Press, Cambridge).
[28]   Berestetskii, V.B., Lifshits, E.M., and Pitaevskii, L.P. (1982), *Relativistic Quantum Theory* (Pergamon, Oxford).
[29]   Bertrand, J., and Bertrand, P. (1987), *Found. Phys.* **17**, 397.

[30]   Bialynicka-Birula, Z., and Bialynicki-Birula, I. (1994), *J. Mod. Opt.* **41**, 2203.

[31]   Bizarro, J.P. (1994), *Phys. Rev. A* **49**, 3255.

[32]   Bjorken, J.D., and Drell, S.D. (1965), *Relativistic Quantum Fields* (McGraw-Hill, New York).

[33]   Böhm, A. (1978), *The Rigged Hilbert Space and Quantum Mechanics* (Springer, Berlin).

[34]   Bohm, D. (1979), *Quantum Theory* (Dover, New York).

[35]   Böhmer, B., and Leonhardt, U. (1995), *Opt. Commun.* **118**, 181.

[36]   Bohr, N. (1935), *Phys. Rev.* **48**, 696.

[37]   Born, M. (1956), *Physics in My Generation* (Pergamon, London).

[38]   Born, M., and Wolf, E. (1970), *Principles of Optics* (Pergamon, Oxford).

[39]   Braunstein, S.L. (1990), *Phys. Rev. A* **42**, 474.

[40]   Braunstein, S.L., Caves, C.M., and Milburn, G.J. (1991), *Phys. Rev. A* **43**, 1153.

[41]   Breitenbach, G., Müller, T., Pereira, S.F., Poizat, J.-Ph., Schiller, S., and Mlynek, J. (1995), *J. Opt. Soc. Am. B* **12**, 2304.

[42]   Brendel, J., Schütrumpf, S., Lange, R., Martienssen W., and Scully, M.O. (1988), *Europhys. Lett.* **5**, 223.

[43]   Brillouin, L. (1926), *Computes Rendus* **183**, 24.

[44]   Brunner, W.H., Paul, H., and Richter, G. (1965), *Ann. Phys.* (Leipzig) **15**, 17.

[45]   Busch, P., Grabowski, M., and Lahti, P.J. (1994), *Phys. Rev. A* **50**, 2881.

[46]   Busch, P., Grabowski, M., and Lahti, P.J. (1995), *Operational Quantum Physics* (Springer, Berlin).

[47]   Bužek, V., and Knight, P.L. (1995), *Prog. Opt.* **34**, 1.

[48]   Bužek, V., Keitel, C.H., and Knight, P.L. (1995), *Phys. Rev. A* **51**, 2575.

[49]   Bužek, V., Keitel, C.H., and Knight, P.L. (1995), *Phys. Rev. A* **51**, 2594.

[50]   Cahill, K.E. (1965), *Phys. Rev.* **138**, 1566.

[51]   Cahill, K.E., and Glauber, R.J. (1969), *Phys. Rev.* **177**, 1857.

[52]   Cahill, K.E., and Glauber, R.J. (1969), *Phys. Rev.* **177**, 1882.

[53]   Campos, R.A., Saleh, B.E.A., and Teich, M.C. (1989), *Phys. Rev. A* **40**, 1371.

[54]   Carmichael, H.J. (1987), *J. Opt. Soc. Am. B* **4**, 1588.

[55]   Carmichael, H.J. (1993), *An Open Systems Approach to Quantum Optics* (Springer, Berlin).

[56]   Caves, C.M. (1982), *Phys. Rev. D* **26**, 1817.

[57]   Cohen, L., and Scully, M.O. (1986), *Found. Phys.* **16**, 295.

[58]   Cohen-Tannoudji, C., Diu, B., and Laloë, F. (1977), *Quantum Mechanics* (Wiley, New York).

[59]   Cohen-Tannoudji, C., Dupont-Roc, J., and Grynberg, G. (1989), *Photons and Atoms* (Wiley, New York).

[60]   Cohen-Tannoudji, C., Dupont-Roc, J., and Grynberg, G. (1992), *Atom-Photon Interactions* (Wiley, New York).

[61]   Cohendet, O., Combe, Ph., Sirugue, M., and Sirugue-Collin, M. (1988), *J. Phys. A* **21**, 2875.

[62]   Cohendet, O., Combe, Ph., and Sirugue-Collin, M. (1990), *J. Phys. A* **23**, 2001.

[63]   Corbett, J.V., and Hurst, C.A. (1978), *J. Austral. Math. Soc. B* **20**, 182.

[64]   Cormack, A.M. (1963), *J. Appl. Phys.* **34**, 2722.

[65]   Courant, R., and Hilbert, D. (1953), *Methods of Mathematical Physics* (Interscience Publishers, New York).

[66]   D'Ariano, G.M., and Paris, M.B.A. (1993), *Phys. Rev. A* **48**, 4039.

[67]   D'Ariano, G.M., and Paris, M.B.A. (1993), *Phys. Rev. A* **49**, 3022.
[68]   D'Ariano, G.M., Macchiavello, C., and Paris, M.B.A. (1994), *Phys. Rev. A* **50**, 4298.
[69]   D'Ariano, G.M., Macchiavello, C., and Paris, M.B.A. (1994), *Phys. Lett. A* **195**, 31.
[70]   D'Ariano, G.M., Leonhardt, U., and Paul, H. (1995), *Phys. Rev. A* **52**, R1801.
[71]   D'Ariano, G.M., Macchiavello, C., and Paris, M.B.A. (1996), *Opt. Commun.* **129**, 6.
[72]   Daudet, H., Deschamps, P., Dion, B., MacGregor, A.D., MacSween, D., McIntyre, R.J., Trottier, C., and Webb, P.P. (1993), *Appl. Opt.* **32**, 3894.
[73]   Davis, L.M., and Parigger, Ch. (1992), *Measurement Science and Technology* **3**, 85.
[74]   Davydov, A.S. (1991), *Quantum Mechanics* (Pergamon, Oxford).
[75]   Dicke, R.H. (1946), *Rev. Sci. Instrum.* **17**, 268.
[76]   Dieks, D. (1982), *Phys. Lett.* **92A**, 271.
[77]   Dirac, P.A.M. (1927), *Proc. Roy. Soc. London* **A114**, 243.
[78]   Dirac, P.A.M. (1984), *The Principles of Quantum Mechanics* (Clarendon, Oxford).
[79]   Dowling, J.P., Schleich, W.P., and Wheeler, J.A. (1991), *Ann. Phys. (Leipzig)* **48**, 423.
[80]   Dowling, J.P., Agarwal, G.S., and Schleich, W.P. (1994), *Phys. Rev. A* **49**, 4101.
[81]   Drummond, P.D., and Gardiner, C.W. (1980), *J. Phys. A* **13**, 2353.
[82]   Dunn, T.J., Walmsley, I.A., and Mukamel, S. (1995), *Phys. Rev. Lett.* **74**, 884.
[83]   Einstein, A. (1905), *Ann. Phys.* **17**, 132.
[84]   Einstein, A., Rosen, N., and Podolsky, B. (1935), *Phys. Rev.* **47**, 777.
[85]   Engen, G.F., and Hoer, C.A. (1972), *IEEE Trans. Instrum. Meas.* **21**, 470.
[86]   Englert, B.-G., and Wódkiewicz, K. (1995), *Phys. Rev. A* **51**, R2661.
[87]   Englert, B.-G., Wódkiewicz, K., and Riegler, P. (1995), *Phys. Rev. A* **52**, 1704.
[88]   Engstrom, R.W. (1980), *Photomultiplier Handbook* (RCA Corporation, Lancaster).
[89]   Erdélyi, A., Magnus, W., Oberhettinger, F., and Tricomi, F.G. (1953), *Higher Transcendental Functions* (McGraw-Hill, New York).
[90]   Fano, U. (1957), *Rev. Mod. Phys.* **29**, 74.
[91]   Faridani, A. (1991), Reconstructing from Efficiently Sampled Data in Parallel-Beam Computed Tomography, in *Inverse Problems and Imaging*, Pitman Research Notes in Mathematics Series, vol. 245, ed. Roach, G.F. (Longman, Harlow).
[92]   Faridani, A. (1996), Results, Old and New, in Computed Tomography, in *The IMA Volumes in Mathematics and its Applications*, ed. Chavent, G., Papanicolaou, G., Sacks, P., and Symes, W. (Springer, Berlin).
[93]   Fearn, H. (1990), *Quantum Opt.* **2**, 113.
[94]   Feenberg, E. (1933), The Scattering of Slow Electrons by Neutral Atoms, Harvard University D. Phil. thesis.
[95]   Feynman, R.P. (1985), *QED: The Strange Theory of Light and Matter* (Princeton University Press, Princeton).
[96]   Freyberger, M., and Schleich, W.P. (1993), *Phys. Rev. A* **47**, R30.
[97]   Freyberger, M., Vogel, K., and Schleich, W.P. (1993), *Phys. Lett. A* **176**, 41.
[98]   Friedrich, H. (1990), *Theoretical Atomic Physics* (Springer, Berlin).

[99]   Gale, W., Guth, E., and Trammell, G.T. (1968), *Phys. Rev.* **165**, 1434.
[100]   Gardiner, C.W. (1991), *Quantum Noise* (Springer, Berlin).
[101]   Gel'fand, I.M., and Shilov, G.E. (1965), *Generalized Functions* (Academic, San Diego).
[102]   Gerchberg, R.W., and Saxton, W.O. (1971), *Optik* **34**, 275.
[103]   Gerchberg, R.W., and Saxton, W.O. (1972), *Optik* **35**, 237.
[104]   Gerhardt, H., Welling, H., and Frölich, (1973), *Appl. Phys.* **2**, 91.
[105]   Gerhardt, H., Büchler, U., and Litfin, G. (1974), *Phys. Lett.* **49A**, 119.
[106]   Glauber, R.J. (1963), *Phys. Rev. Lett.* **10**, 84.
[107]   Goethe, J.W. (1982), *Faust. Der Tragödie erster Teil* (Reclam, Leipzig).
[108]   Haken, H. (1981), *Light* (Elsevier, Amsterdam).
[109]   Harris, E.G. (1972), *A Pedestrian Approach to Quantum Field Theory* (Wiley, New York).
[110]   Heisenberg, W. (1969), *Der Teil und das Ganze* (Piper, München). English translation: Heisenberg, W. (1971), *Physics and Beyond; Encounters and Conversations* (Harper and Row, New York).
[111]   Helstrom, C.W. (1976), *Quantum Detection and Estimation Theory* (Academic, New York).
[112]   Heritage, J.P., Weiner, A.M., and Thurston, R.N. (1985), *Opt. Lett.* **10**, 609.
[113]   Herman, G.T. (1980), *Image Reconstruction from Projections: The Fundamentals of Computerized Tomography* (Academic, New York).
[114]   Hertz, H. (1887), *Ann. Phys.* **31**, 982.
[115]   Herzog, U. (1996), *Phys. Rev. A* **53**, 1245.
[116]   Hillery, M., O'Connell, R.F., Scully, M.O., and Wigner, E.P. (1984), *Phys. Rep.* **106**, 121 (1984).
[117]   Hong, C.K., and Mandel, L. (1986), *Phys. Rev. Lett.* **56**, 58.
[118]   Hong, C.K., Ou, Z.Y., and Mandel, L. (1987), *Phys. Rev. Lett.* **59**, 2044.
[119]   Hudson, R.L. (1974), *Rep. Math. Phys.* **6**, 249.
[120]   Husimi, K. (1940), *Proc. Phys. Math. Soc. Japan* **22**, 264.
[121]   Huttner, B., and Ben-Aryeh, Y. (1988), *Phys. Rev. A* **38**, 204.
[122]   Itzykson, C., and Zuber, J.B. (1980), *Quantum Field Theory* (McGraw-Hill, New York).
[123]   Ivanović, I.D. (1981), *J. Phys. A* **14**, 3241.
[124]   Ivanović, I.D. (1983), *J. Math. Phys.* **24**, 1199.
[125]   Jammer, M. (1989), *The Conceptual Development of Quantum Mechanics* (McGraw-Hill, New York).
[126]   Janicke, U., and Wilkens, M. (1995), *J. Mod. Opt.* **42**, 2183.
[127]   Janszky, J., Sibilia, C., Bertolotti, M., and Yushin, Y. (1988), *J. Mod. Opt.* **35**, 1757.
[128]   Janszky, J., Sibilia, C., and Bertolotti, M. (1991), *J. Mod. Opt.* **38**, 2467.
[129]   Janszky, J., Adam, P., Sibilia, C., and Bertolotti, M. (1992), *Quantum Opt.* **4**, 163.
[130]   Janszky, J., Kim, M.G., and Kim, M.S. (1996), *Phys. Rev. A* **53**, 502.
[131]   Jex, I., Stenholm, S., and Zeilinger, A. (1995), *Opt. Commun.* **117**, 95.
[132]   Jordan, P. (1935), *Z. Phys.* **94**, 531.
[133]   Kemble, E.C. (1937), *Fundamental Principles of Quantum Mechanics* (Dover, New York).
[134]   Kilin, S.Ya., and Horoshko, D.B. (1995), *Phys. Rev. Lett.* **74**, 5206.
[135]   Kim, Y.S., and Noz, M.E. (1991), *Phase Space Picture of Quantum Mechanics: Group-Theoretical Approach* (World Scientific, Singapore).
[136]   Kiss, T., Herzog, U., and Leonhardt, U. (1995), *Phys. Rev. A* **52**, 2433.

[137]   Klauder, J.R. (1966), *Phys. Rev. Lett.* **16**, 534.
[138]   Klauder, J.R., and Skagerstam, B.-S. (1985), *Coherent States* (World Scientific, Singapore).
[139]   Kramers, H.A. (1926), *Z. Phys.* **39**, 828.
[140]   Kühn, H., Welsch, D.-G., and Vogel, W. (1994), *J. Mod. Opt.* **41**, 1607.
[141]   Kühn, H., Welsch, D.-G., and Vogel, W. (1995), *Phys. Rev. A* **51**, 4240.
[142]   Kwiat, P.G., Steinberg, A.M., Chiao, R.Y., Eberhard, P.H., and Petroff, M.D. (1993), *Phys. Rev. A* **48**, R867.
[143]   Lai, W.K., Bužek, V., and Knight, P.L. (1991), *Phys. Rev. A* **43**, 6323.
[144]   Lalović, D., Davidović, D.M., and Bijedić, N. (1992), *Phys. Rev. A* **46**, 1206.
[145]   Landau, L.D., and Lifshitz, E.M. (1977), *Quantum Mechanics* (Pergamon, Oxford).
[146]   Lange, R., Brendel, J., Mohler, E., and Martienssen, W. (1988), *Europhys. Lett.* **5**, 619.
[147]   Larsen, U. (1990), *J. Phys. A* **23**, 1041.
[148]   Lee, Ch.T. (1994), *Phys. Rev. A* **48**, 2285.
[149]   Lee, H.W. (1995), *Phys. Rep.* **259**, 147.
[150]   Lehner, J., Leonhardt, U., and Paul, H. (1996), *Phys. Rev. A* **53**, 2727.
[151]   Leonhardt, U. (1993), Quantum Theory of Simple Optical Instruments, Humboldt University D. Phil. thesis.
[152]   Leonhardt, U., and Paul, H. (1993), *Phys. Rev. A* **47**, R2460.
[153]   Leonhardt, U., and Paul, H. (1993), *J. Mod. Opt.* **40**, 1745.
[154]   Leonhardt, U., and Paul, H. (1993), *Phys. Scr.* **T48**, 45.
[155]   Leonhardt, U. (1993), *Phys. Rev. A* **48**, 3265.
[156]   Leonhardt, U., and Paul, H. (1993), *Phys. Rev. A* **48**, 4598.
[157]   Leonhardt, U. (1994), *Phys. Rev. A* **49**, 1231.
[158]   Leonhardt, U., and Jex, I. (1994), *Phys. Rev. A* **49**, R1555.
[159]   Leonhardt, U., and Paul, H. (1994), *Phys. Rev. Lett.* **72**, 4086.
[160]   Leonhardt, U., and Paul, H. (1994), *J. Mod. Opt.* **41**, 1427.
[161]   Leonhardt, U., and Paul, H. (1994), *Phys. Lett. A* **193**, 117.
[162]   Leonhardt, U., Vaccaro, J.A., Böhmer, B., and Paul, H. (1995), *Phys. Rev. A* **51**, 84.
[163]   Leonhardt, U., and Paul, H. (1995), *Progr. Quantum Electron.* **19**, 89.
[164]   Leonhardt, U., and Vaccaro, J.A. (1995), *J. Mod. Opt.* **42**, 939.
[165]   Leonhardt, U. (1995), *Phys. Rev. Lett.* **74**, 4101.
[166]   Leonhardt, U., Böhmer, B., and Paul, H. (1995), *Opt. Commun.* **119**, 296.
[167]   Leonhardt, U., Paul, H., and D'Ariano, G.M. (1995), *Phys. Rev. A* **52**, 4899.
[168]   Leonhardt, U., and Raymer, M.G. (1996), *Phys. Rev. Lett.* **76**, 1989.
[169]   Leonhardt, U., Munroe, M., Kiss, T., Richter, Th., and Raymer, M.G. (1996), *Opt. Commun.* **127**, 144.
[170]   Leonhardt, U. (1996), *Phys. Rev. A* **53**, 2998.
[171]   Leonhardt, U. (1996), *Acta Physica Slovaca* **46**, 309.
[172]   Leonhardt, U., and Munroe, M. (1996), *Phys. Rev. A* **54**, 3682.
[173]   Leubner, C. (1985), *Eur. J. Phys.* **6**, 299.
[174]   Levenson, J.A., Abram, I., Rivera, T., and Fayolle, P. (1993), *Phys. Rev. Lett.* **70**, 267.
[175]   Lieb, E.H. (1990), *J. Math. Phys.* **31**, 594.
[176]   Lohmann, A.W. (1993), *J. Opt. Soc. Am. A* **10**, 2181.
[177]   London, F. (1927), *Z. Phys.* **40**, 193.
[178]   Loudon, R. (1983), *The Quantum Theory of Light* (Clarendon Press, Oxford).

[179] Loudon, R., and Knight, P.L. (1987), *J. Mod. Opt.* **34**, 709.
[180] Louisell, W.H. (1963), *Phys. Lett.* **7**, 60.
[181] Louisell, W.H. (1973), *Quantum Statistical Properties of Radiation* (Wiley, New York).
[182] Luis, A., and Sánchez Soto, L.L. (1995), *Quantum Opt.* **7**, 153.
[183] Lukš, A., and Peřinová, V. (1993), *Phys. Scr.* **T48**, 94.
[184] Lukš, A., Peřinová, V., and Křepelka, J. (1994), *Phys. Rev. A* **50**, 818.
[185] Lütkenhaus, N., and Barnett, S.M. (1995), *Phys. Rev. A* **51**, 3340.
[186] Lynch, R. (1995), *Phys. Rep.* **256**, 367.
[187] Mandel, L., and Wolf, E. (1995), *Optical Coherence and Quantum Optics* (Cambridge University Press, Cambridge).
[188] McAlister, D.F., Beck, M., Clarke, C., Mayer, A., and Raymer, M.G. (1995), *Opt. Lett.* **20**, 1181.
[189] Mertz, L. (1988), *Appl. Opt.* **27**, 3429.
[190] Meystre, P., and Sargent, M. III (1991), *Elements of Quantum Optics* (Springer, Berlin).
[191] Moyal, J.E. (1949), *Proc. Cambridge Philos. Soc.* **45**, 99.
[192] Mukunda, N. (1979), *Am. J. Phys.* **47**, 182.
[193] Munroe, M., Boggavarapu, D., Anderson, M.E., and Raymer, M.G. (1995), *Phys. Rev. A* **52**, R924.
[194] Natterer, F. (1986), *The Mathematics of Computerized Tomography* (Wiley, Chichester).
[195] Newton, R.G., and Young, B.-L. (1968), *Ann. Phys.* (New York) **49**, 393.
[196] Newton, R.G. (1980), *Ann. Phys.* (New York) **124**, 327.
[197] Noh, J.W., Fougères, A., and Mandel, L. (1991), *Phys. Rev. Lett.* **67**, 1426.
[198] Noh, J.W., Fougères, A., and Mandel, L. (1992), *Phys. Rev. A* **45**, 424.
[199] Noh, J.W., Fougères, A., and Mandel, L. (1992), *Phys. Rev. A* **46**, 2840.
[200] Noh, J.W., Fougères, A., and Mandel, L. (1993), *Phys. Scr.* **T48**, 29.
[201] Noh, J.W., Fougères, A., and Mandel, L. (1993), *Phys. Rev. Lett.* **71**, 2579.
[202] Nussenzveig, H.M. (1974), *Introduction to Quantum Optics* (Gordon and Breach, New York).
[203] Opatrný, T., Bužek, V., Bajer, J., and Drobný, G. (1995), *Phys. Rev. A* **52**, 2419.
[204] Opatrný, T., Welsch, D.-G., and Bužek, V. (1996), *Phys. Rev. A* **53**, 3822.
[205] Orlowski, A., and Paul, H. (1994), *Phys. Rev. A* **50**, R921.
[206] Ou, Z.Y., Hong, C.K., and Mandel, L. (1987), *Opt. Commun.* **63**, 118.
[207] Ou, Z.Y., Pereira, S.F., Kimble, H.J., and Peng, K.C. (1992), *Phys. Rev. Lett.* **68**, 3663.
[208] Ou, Z.Y., and Kimble, H.J. (1995), *Phys. Rev. A* **52**, 3126.
[209] Paul, H., Brunner, W.H., and Richter, G. (1966), *Ann. Phys.* (Leipzig) **17**, 262.
[210] Paul, H. (1974), *Fortschr. Phys.* **22**, 657.
[211] Paul, H. (1982), *Rev. Mod. Phys.* **54**, 1061.
[212] Pauli, W. (1933), Die allgemeinen Prinzipien der Wellenmechanik, in *Handbuch der Physik*, ed. Geiger, H., and Scheel, K. (Springer, Berlin). English translation: Pauli, W. (1980), *General Principles of Quantum Mechanics* (Springer, Berlin).
[213] Pegg, D.T., and Barnett, S.M. (1988), *Europhys. Lett.* **6**, 483.
[214] Pegg, D.T., and Barnett, S.M. (1989), *Phys. Rev. A* **39**, 1665.
[215] Pegg, D.T., Vaccaro, J.A., and Barnett, S.M. (1994), Quantum Optical Phase and its Measurement, in *Quantum Optics VI*, ed. Harvey, J.D., and Walls, D.F. (Springer, Berlin).

[216]   Perelomov, A.M. (1986), *Generalized Coherent States and Their Applications* (Springer, Berlin).

[217]   Peřina, J. (1991), *Quantum Statistics of Linear and Nonlinear Optical Phenomena* (Kluwer, Dortrecht).

[218]   Peřinová, V., Lukš, A., Křepelka, J., Sibilia, C., and Bertolotti, M. (1991), *J. Mod. Opt.* **38**, 2429.

[219]   Πλάτωνος Πολιτεία. English translation: Plato (1935), *Republic*, Book VII, p. 514, in *The Loeb Classical Library* **L276**, Vol. VI (Harvard University Press, Cambridge MA).

[220]   Poincaré, H. (1956), Chance, in *The World of Mathematics*, ed. Newman, J.R. (Simon and Schuster, New York).

[221]   Polzik, E.S., Carry, J., and Kimble, H.J. (1992), *Phys. Rev. Lett.* **68**, 3020.

[222]   Popper, K.R. (1982), *Quantum Theory and the Schism in Physics* (Hutchinson, London).

[223]   Prasad, S., Scully, M.O., and Martienssen, W. (1987), *Opt. Commun.* **62**, 139.

[224]   Proakis, J.G., Rader, C.M., Ling, F., and Nikias, C.L. (1992), *Advanced Digital Signal Processing* (Macmillan, New York).

[225]   Prudnikov, A.P., Brychkov, Yu.A., and Marichev, O.I. (1992), *Integrals and Series* (Gordon and Breach, New York).

[226]   Radon, J. (1917), *Berichte über die Verhandlungen der Königlich-Sächsischen Gesellschaft der Wissenschaften zu Leipzig, Mathematisch-Physische Klasse* **69**, 262.

[227]   Raymer, M.G., Beck, M., and McAlister, D.F. (1994), *Phys. Rev. Lett.* **72**, 1137.

[228]   Raymer, M.G. (1994), *Am. J. Phys.* **62**, 986.

[229]   Raymer, M.G., Cooper, J., Carmichael, H.J., Beck, M., and Smithey, D.T. (1995), *J. Opt. Soc. Am. B* **12**, 1801.

[230]   Reck, M., Zeilinger, A., Bernstein, H.J., and Bertani, P. (1994), *Phys. Rev. Lett.* **73**, 58.

[231]   Reid, M.D., and Walls, D.F. (1984), *Phys. Rev. Lett.* **53**, 955.

[232]   Reid, M.D., and Drummond, P.D. (1988), *Phys. Rev. Lett.* **60**, 2731.

[233]   Richter, Th. (1996), *Phys. Lett. A* **211**, 327.

[234]   Richter, Th., and Wünsche, A. (1996), *Phys. Rev. A* **53**, R1974.

[235]   Richter, Th., and Wünsche, A. (1996), *Acta Physica Slovaca* **46**, 487.

[236]   Robertson, H.P. (1929), *Phys. Rev.* **34**, 163.

[237]   Riegler, P., and Wódkiewicz, K. (1994), *Phys. Rev. A* **49**, 1387.

[238]   Royer, A. (1985), *Phys. Rev. Lett.* **55**, 2745.

[239]   Royer, A. (1989), *Found. Phys.* **19**, 3.

[240]   Schiller, S., Breitenbach, G., Pereira, S.F., Müller, T., and Mlynek, J. (1996), *Phys. Rev. Lett.* **77**, 2933.

[241]   Schleich, W.P., Walther, H., and Wheeler, J.A. (1988), *Found. Phys.* **18**, 953.

[242]   Schleich, W.P., Horowicz, R.J., and Varro, S. (1989), A Bifurcation in Squeezed State Physics: But How? or Area-of-Overlap in Phase Space as a Guide to the Phase Distribution and the Action-Angle Wigner Function of a Squeezed State, in *Quantum Optics V*, ed. Harvey, J.D., and Walls, D.F. (Springer, Berlin).

[243]   Schleich, W.P., Pernigo, M., and Fam Le Kien (1991), *Phys. Rev. A* **44**, 2172.

[244]   Schleich, W.P., Bandilla, A., and Paul, H. (1992), *Phys. Rev. A* **45**, 6652.

[245]   Schleich, W.P., Krähmer, D.S., and Mayr, E. (1997), *Quantum Optics in Phase Space* (VCH, Weinheim).

[246] Schrödinger, E. (1935), *Naturwissenschaften* **23**, 807.

[247] Schwinger, J. (1952), *U.S. Atomic Energy Commission Report No. NYO-3071* (U.S. GPO, Washington, D.C.); reprinted in Biederharn, L.C., and van Dam, H. (1965), *Quantum Theory of Angular Momenta* (Academic, New York).

[248] Scully, M.O. (1983), *Phys. Rev. D* **28**, 2477.

[249] Scully, M.O., and Wódkiewicz, K. (1994), *Found. Phys.* **24**, 85.

[250] Scully, M.O., Walther, H., and Schleich, W.P. (1994), *Phys. Rev. A* **49**, 1562.

[251] Scully, M.O., and Zubairy, M.S. (1997), *Quantum Optics* (Cambridge University Press, Cambridge).

[252] Shapiro, J.H., Yuen, H.P., and Machado Mata, J.A. (1979), *IEEE Trans. Inf. Theory* **IT-25**, 179.

[253] Shapiro, J.H., and Wagner, S.S. (1984), *IEEE J. Quantum Electron.* **QE-20**, 803.

[254] Shapiro, J.H., and Shepard, S.R. (1991), *Phys. Rev. A* **43**, 3795.

[255] Smithey, D.T., Beck, M., Raymer, M.G., and Faridani, A. (1993), *Phys. Rev. Lett.* **70**, 1244.

[256] Smithey, D.T., Beck, M., Cooper, J., Raymer, M.G., and Faridani, A. (1993), *Phys. Scr.* **T48**, 35.

[257] Smithey, D.T., Beck, M., Cooper, J., and Raymer, M.G. (1993), *Phys. Rev. A* **48**, 3159.

[258] Snyder, A.W., and Love, J.D. (1991), *Optical Waveguide Theory* (Science Paperbacks, London).

[259] Stenholm, S. (1992), *Ann. Phys.* (New York) **218**, 233.

[260] Stenholm, S. (1995), *Appl. Phys. B* **60**, 243.

[261] Sudarshan, E.C.G. (1963), *Phys. Rev. Lett.* **10**, 277.

[262] Szegö, G. (1975), *Orthogonal Polynomials* (American Mathematical Society, Providence).

[263] Tatarskii, V.I. (1983), *Sov. Phys. Usp.* **26**, 311.

[264] Titchmarsh, E.C. (1937), *Introduction to the Theory of Fourier Integrals* (Clarendon, Oxford), Chap. V.

[265] Toll, J.S. (1956), *Phys. Rev.* **104**, 1760.

[266] Torgerson, J.R., and Mandel, L. (1996), *Phys. Rev. Lett.* **76**, 3939.

[267] Törmä, P., Stenholm, S., and Jex, I. (1995), *Phys. Rev. A* **52**, 4812.

[268] Törmä, P., Stenholm, S., and Jex, I. (1995), *Phys. Rev. A* **52**, 4853.

[269] Törmä, P., Jex, I., and Stenholm, S. (1996), *J. Mod. Opt.* **43**, 245.

[270] Umezawa, H., Matsumoto, H., and Tachiki, M. (1982), *Thermo Field Dynamics and Condensed States* (Elsevier, Amsterdam).

[271] Vacarro, J.A., and Pegg, D.T. (1990), *Phys. Rev. A* **41**, 5156.

[272] Vaccaro, J.A. (1995), *Opt. Commun.* **113**, 421.

[273] Vaccaro, J.A. (1995), *Phys. Rev. A* **51**, 3309.

[274] Vaccaro, J.A. (1995), *Phys. Rev. A* **52**, 3474.

[275] Vaccaro, J.A., and Barnett, S.M. (1995), *J. Mod. Opt.* **42**, 2165.

[276] van der Plank, R.W.F., and Suttorp, L.G. (1994), *Opt. Commun.* **112**, 145.

[277] Várilly, J.C., and Gracia-Bondia, J.M. (1989), *Ann. Phys.* (New York) **190**, 107.

[278] Vogel, K., and Risken, H. (1989), *Phys. Rev. A* **40**, 2847.

[279] Vogel, K., and Schleich, W.P. (1992), More on Interference in Phase Space, in *Lectures delivered at Les Houches, Session LIII, Systèmes Fondamentaux en Optique Quantique* (Elsevier, Amsterdam).

[280] Vogel, W., and Grabow, J. (1993), *Phys. Rev. A* **47**, 4227.

[281]   Vogel, W., and Welsch, D.-G. (1994), *Lectures on Quantum Optics* (Akademie-Verlag, Berlin).
[282]   Walker, N.G., and Carroll, J.E. (1984), *Electron. Lett.* **20**, 981.
[283]   Walker, N.G. (1987), *J. Mod. Opt.* **34**, 15.
[284]   Walls, D.F., and Milburn, G.J. (1994), *Quantum Optics* (Springer, Berlin).
[285]   Wefers, M., and Nelson, K. (1995), *J. Opt. Soc. Am. B* **12**, 1343.
[286]   Wefers, M., and Nelson, K. (1995), *Opt. Lett.* **20**, 1047.
[287]   Weigert, S. (1996), *Phys. Rev. A* **53**, 2078.
[288]   Wentzel, G. (1926), *Z. Phys.* **38**, 518.
[289]   Weyl, H. (1950), *The Theory of Groups and Quantum Mechanics* (Dover, New York).
[290]   Wigner, E.P. (1932), *Phys. Rev.* **40**, 749.
[291]   Wilkens, M., and Meystre, P., (1991), *Phys. Rev. A* **43**, 3832.
[292]   Wiseman, H.J. (1995), *Phys. Rev. Lett.* **75**, 4587.
[293]   Wódkiewicz, K. (1984), *Phys. Rev. Lett.* **52**, 1064.
[294]   Wódkiewicz, K. (1984), *Phys. Lett. A* **115**, 304.
[295]   Wódkiewicz, K. (1995), *Phys. Rev. A* **51**, 2785.
[296]   Wootters, W.K., and Zurek, W.H. (1982), *Nature* **299**, 802.
[297]   Wootters, W.K. (1986), *Found. Phys.* **16**, 391.
[298]   Wootters, W.K. (1987), *Ann. Phys.* (New York) **176**, 1.
[299]   Wootters, W.K., and Fields, B.D. (1989), *Ann. Phys.* (New York) **191**, 363.
[300]   Wünsche, A. (1996), *Quantum Semiclass. Opt.* **8**, 343.
[301]   Yuen, H.P., and Lax, M. (1973), *IEEE Trans. Inf. Theory* **IT-19**, 740.
[302]   Yuen, H.P., and Chan, V.W.S. (1983), *Opt. Lett.* **8**, 177.
[303]   Yuen, H.P., and Shapiro, J.H. (1978), Quantum Statistics of Homodyne and Heterodyne Detection, in *Coherence and Quantum Optics IV*, ed. Mandel, L., and Wolf, E. (Plenum, New York).
[304]   Yuen, H.P., and Shapiro, J.H. (1978), *IEEE Trans. Inf. Theory* **IT-24**, 657.
[305]   Yuen, H.P., and Shapiro, J.H. (1980), *IEEE Trans. Inf. Theory* **IT-26**, 78.
[306]   Zeilinger, A. (1981), *Am. J. Phys.* **49**, 882.
[307]   Zucchetti, A., Vogel, W., Tasche, M., and Welsch, D.-G. (1996), *Phys. Rev. A* **54**, 856.
[308]   Zurek, W.H. (1991), *Physics Today* (October), 36.

Index